ROBERT MANNING
STROZIER LIBRARY

SEP 2 1994

Tallahassee, Florida

ADOLESCENT STORM AND STRESS
An Evaluation of the
Mead–Freeman Controversy

RESEARCH MONOGRAPHS IN ADOLESCENCE
Nancy L. Galambos/Nancy A. Busch-Rossnagel, Editors

Côté • Adolescent Storm and Stress: An Evaluation of the Mead–Freeman Controversy

ADOLESCENT STORM AND STRESS
An Evaluation of the Mead–Freeman Controversy

James E. Côté
University of Western Ontario

LAWRENCE ERLBAUM ASSOCIATES, PUBLISHERS
1994   Hillsdale, New Jersey   Hove, UK

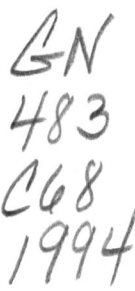

GN
483
C68
1994

Copyright © 1994 by Lawrence Erlbaum Associates, Inc.
All rights reserved. No part of this book may be reproduced in
any form, by photostat, microfilm, retrieval system, or any other
means, without the prior written permission of the publisher.

Lawrence Erlbaum Associates, Inc., Publishers
365 Broadway
Hillsdale, New Jersey 07642

Cover design by Kate Dusza

Library of Congress Cataloging-in-Publication Data

Côté, James E.
    Adolescent storm and stress : an evaluation of the Mead-Freeman controversy / James E. Côté.
        p.   cm. --(Research monographs in adolescence)
    Includes bibliographical references and indexes.
    ISBN 0-8058-1506-6 (cloth)
    1. Adolescence. 2. Adolescent psychology. 3. Nature and nurture.
4. Mead, Margaret, 1901-1978. 5. Freeman, Derek. I. Title.
II. Series.
GN483.C68   1994
305.23'5--dc20
                                                    93-45835
                                                    CIP

Books published by Lawrence Erlbaum Associates are printed on acid-free
paper, and their bindings are chosen for strength and durability.

Printed in the United States of America
10 9 8 7 6 5 4 3 2 1

This book is dedicated
to the young people
currently struggling to
"come of age in Samoa."

A firm knowledge of their history is particularly critical for young people who must make their life choices in the context of the present clash between tradition and a form of modernity that . . . negates and undermines the authentic values of their past. . . . [A] firm knowledge of the past choices made by their forbears . . . [is] enormously important if young people are to maintain a sure sense of identity as they face the difficult constraints and choices of the present.
—Leacock (1987, p. 180)

Contents

Series Foreword   ix

Preface   xi

1   The Mead–Freeman Controversy: Mead on Trial   1

    The Context of Mead's Research   *1*
    The Controversy Sparked by Freeman's Book:
       Mead on Trial   *3*
    The Central Figures in the Controversy   *4*
    Methodologies Appropriate to the Evaluation of
       Mead's Coming-of-Age Thesis   *11*

2   Freeman's Case Against Mead   14

    The Premise of Freeman's Case: Adolescent
       Biology   *14*
    The Politics of the Controversy   *19*
    Freeman's Evidence: Adolescent Storm and Stress   *30*
    Lies, Damned Lies, and Statistics   *43*
    Conclusion   *46*

3   Mead's Culpability   48

    Expert Witnesses   *48*
    Material Evidence   *53*
    A Verdict on Mead's Coming-of-Age Thesis   *62*

4 A Social History of Adolescence in Samoa: Precontact Culture   65

Coming of Age in Precontact Samoa   *66*
Analysis   *72*
Sexual Practices in Precontact Samoa   *74*

5 A Social History of Adolescence in Samoa: Changes in Samoan Culture   83

Missionaries and Their Impact   *83*
Missionary Impact on Coming of Age   *87*
Missionary Impact on Sexual Practices   *95*

6 Mead's Samoa   100

Is it Plausible in Terms of the Historical Evidence?   *100*

7 Coming of Age in Contemporary Samoa   122

Western Influence and the Cultural Disenfranchisement of Samoan Youth   *122*
Coming of Age in Western Samoa, 1990   *125*
Dealing With an Unfolding Tragedy Facing Western Samoan Youth   *136*
The Cultural Heritage of the Young: Independence or Dependence?   *145*

8 Conclusion: Mead's Samoa in Sociological Perspective   149

Adolescence as a Stage of Life   *149*
Institutionalized Moratoria and Mead's Samoa   *152*
The Issue of Choice   *160*
Limitations of the Present Study   *164*
Future Research   *165*

References   170

Author Index   178

Subject Index   182

Series Foreword

Research Monographs in Adolescence is a series of theoretically grounded empirical works in the form of authored monographs. Drawing from the knowledge and expertise of scholars in multiple disciplines, such as psychology, sociology, and anthropology, the books in this series are written by leaders in the field for their peers in the academic community. These original and state-of-the-art works will be useful as archival material and will serve as valuable resources for upper level undergraduate and graduate courses. The monographs will not only advance our understanding of adolescence through the finest research possible, but will provide information necessary to design useful intervention strategies and to direct social policy.

The premier book in this series, *Adolescent Storm and Stress: An Evaluation of the Mead–Freeman Controversy*, by sociologist James Côté, is an excellent example of the scholarship to which this series aspires. By providing a compelling and careful examination of critical arguments in the Mead–Freeman controversy, this volume makes a significant contribution to our understanding of adolescent development and the transition to adulthood, while shedding light on the politics of the debate.

Nancy L. Galambos
Nancy A. Busch-Rossnagel

Preface

In 1928, Margaret Mead published her first book, entitled *Coming of Age in Samoa*, in which she described to the Western world an exotic culture where people passed through adolescence with a minimum of difficulties. *Coming of Age* brought her instant acclaim and stands as the "most widely read anthropology book ever published," having sold millions of copies in 16 languages (Muuss, 1988, p. 139). This book also launched her career as the best-known and most outspoken anthropologist of the century.

For more than 50 years, images from Mead's pioneering book permeated Western public consciousness of life in so-called simpler societies. Among the public, an image apparently prevailed that Samoa was a veritable paradise, where casual attitudes marked most social relationships, including sexual ones. In addition, Mead's conclusion that Samoan adolescents experienced little "storm and stress" was widely accepted in academic circles.

In 1983, Derek Freeman, an Australian anthropologist, published a book in which he systematically attacked Mead's conclusions about that culture and the way people came of age. Aside from attacking the most famous anthropologist of the century, Freeman took on the cultural anthropology "establishment" by questioning the integrity of its role in the "great debate" of the 20th century—the perennial "nature–nurture" debate. Since then, Freeman's book, entitled *Margaret Mead and Samoa: The Making and Unmaking of an Anthropological Myth*, has received so much attention that one observer remarked that Freeman had sparked "the biggest debate in the social sciences for years" (Crocombe, 1989, p. 38).

On the one hand, one could dismiss the whole matter as that of one person's word against another's. On the other hand, there are important implications to this controversy. Most generally at stake is the nature–nurture debate itself, and the presumption that Freeman has scored a victory for the "nature" side. More specifically at stake are assumptions about the period we call "adolescence and youth," and the belief that adolescence is inevitably a period of "storm and stress"—of biological/hormonal disequilibrium. But, perhaps most important are the implications of this controversy for young people in all societies, including contemporary Samoa. I refer here to the social expectations and economic policies that affect their well-being and life chances. If these expectations and policies are to be fair, they must be based on sound knowledge about the intrinsic nature of adolescence, and not on hastily drawn conclusions clouded by territorial academic disputes. For these reasons, I set out to carefully analyze the questions raised by Freeman and evaluate the charges he has leveled against Mead.

When I first began investigating the Mead–Freeman controversy, I had no intention of writing such a detailed manuscript on the matter. However, I did not anticipate the complexities involved in sorting out the many claims made in the affair. As I was sorting things out, however, for the reasons just stated, I became convinced that a book had to be written to help other interested parties evaluate these claims. To keep my investigation from being sidetracked by the complexities of the Mead–Freeman controversy, I remained focused on Mead's original research question. This question was whether coming of age was easier in Samoa than in the United States during the 1920s. I refer to it as Mead's "coming-of-age thesis." In assessing this thesis, two issues must be settled: To what extent are her various *conclusions* plausible? To what extent were her *observations* accurate?

The evidence I accumulated to tackle these issues is presented throughout the eight chapters of this book, each of which assesses Freeman's criticisms from a different perspective.

Chapters 1 through 3 present a comprehensive evaluation of the charges made by Freeman regarding Mead's coming-of-age conclusions. Over the span of these three chapters, the scientific quality of Freeman's charges is assessed against the plausibility of Mead's coming-of-age thesis. As we see, when the pertinent evidence is examined, very few of Freeman's charges persuade us to reject Mead's thesis.

In chapters 4 through 6, the accuracy of Mead's coming-of-age observations is evaluated with a sociohistorical analysis of Samoan culture. Freeman's case against Mead hinged on his assertion that she was largely wrong in her contentions about adolescence because she had a limited and biased perspective regarding Samoan culture. He claimed that he has a more thorough and accurate knowledge of the culture. In order to evaluate this crucial aspect of Freeman's critique, we undertake a sociohistorical analysis that traces the trajectory of social change initiated by missionaries in the 1800s (chapters 4 and 5). Based on written documents, this analysis allows us to determine if we can pinpoint the "Samoa" that Mead recorded in the 1920s (chapter 6). Again, we find little written evidence that leads us to seriously question the validity of Mead's observations.

Chapter 7 rounds out the sociohistorical evaluation of Mead's coming-of-age thesis by providing a glimpse at the "coming of age" experienced by contemporary Samoan youth. When we compare what prevailed in "precontact" Samoan culture with the circumstances confronting Samoan youth today, we find an insidious and inexorable disenfranchisement of young people there. What was once a place where most young people apparently came of age with relative ease, has become a place where many young people experience great difficulty finding a place in their society. This problem is now so serious that Western Samoa has one of the highest youth suicide rates in the world. But, Freeman denied the relevance of social change for the way Samoans come of age, claiming that adolescence has been stressful since before contact with the West. By doing so, he overlooked the possibility that much of what he witnessed during the 1960s (when he did most of his research) was a product of social change that had taken place since the 1920s. Thus, the missing component in this aspect of the controversy is the social change that was set in motion by the missionaries during the 1800s. And, herein lies the crux of the controversy—it is only within its context that we can understand the controversy itself, and the context is the social change and cultural disenfranchisement that has altered the way Samoans come of age. In light of what has happened as a result of contact with the West, this chapter ends with a set of policy recommendations regarding what might be done to rectify the cultural disenfranchisement of Samoan youth.

The concluding chapter (chapter 8) completes the interdisciplinary evaluation of Mead's coming-of-age thesis by examining from a sociological perspective the "Samoa" Mead described, and evaluating her claim that it would have provided an ease of passage from childhood to adulthood. We see here how her pioneering study introduced questions that are still pertinent to the study of adolescence. Although crucial elements of the Mead–Freeman controversy as they pertain to Mead's central thesis are resolved in this chapter, several questions are raised as to what is happening to young people in contemporary Samoa and what will happen there in the future (e.g., on the issue of "choosing" one's identity). The chapter ends with a set of recommendations for future research that will help us to continue to address these questions, both in terms of understanding the "adolescent process" and in terms of comprehending the social disorganization that young people in places like Samoa now face as a result of Western influence.

All told, my investigation of the Mead–Freeman controversy led me to the conclusion that Mead's coming-of-age thesis is quite plausible when viewed from a variety of perspectives, and when viewed in the context of the "Samoa" she studied. Yes, there are problems with some of what she wrote in *Coming of Age*. But there is little reason to believe that she was wrong in most of what she reported—contrary to what Freeman claimed and despite the mythology surrounding her book. With respect to my ability to arrive at a reasonable conclusion in this matter, I believe the fact that I am an "outsider" to both the anthropological community and Samoan culture has helped me to see the controversy in a way that provides a unique and useful contribution to the literature on Samoa: As a sociologist, I am not bound by an allegiance to the anthropological community; and as a non-Samoan I am not bound by a code that prevents me from discussing what some see as the less flattering aspects of Samoan culture. Instead, I am free to apply my expertise unrestrained by the partisanships that have marked some attempts to make sense out of Mead's work in Samoa and Freeman's criticism of her work.

ACKNOWLEDGMENTS

I would like to thank my close colleagues, Anton Allahar, Charles Levine, and Benjamin Singer, for their helpful comments on a previous draft of this manuscript as well as for their continuing personal support. I am also indebted to Dean Emoke Szathmary for her encouragement from the beginning of this project, and to Gerald Adams for his ongoing sound advice. In addition, it was a delight to work with the editors of the Research Monographs in Adolescence series, Nancy L. Galambos and Nancy A. Busch-Rossnagel. I appreciate their sharp insights and competent guidance. Finally, the assistance of the J. B. Smallman Publication Fund, Faculty of Social Science, The University of Western Ontario, is acknowledged.

James E. Côté

CHAPTER ONE

The Mead–Freeman Controversy: Mead on Trial

THE CONTEXT OF MEAD'S RESEARCH

Mead undertook her study during a period when it was widely believed that "adolescence" was associated with a period of turmoil caused by biological instabilities stimulated at puberty. As Freeman (1983) has noted, this belief was a cornerstone of the "nature" side of the nature–nurture debate. An implication of this position is that adolescence is an "affliction" that inevitably occurs regardless of social or cultural circumstances. Accordingly, it was argued that it does not matter what experiences and opportunities are available to the young, all pass through a period of what was dubbed *storm and stress* (from the German *Sturm und Drang*). In the introductory chapter to *Coming of Age in Samoa* Mead (1928) described the climate that prevailed in the 1920s:

> a great mass of theorizing about adolescence is flooding the book shops; so the psychologist in America tried to account for the restlessness of youth. The result was works like that of Stanley Hall of "Adolescence," which ascribed to the period through which the children were passing, the cause of their own conflict and distress. Adolescence was characterized as the period in which idealism flowered and rebellion against authority waxed strong, a period during which difficulties and conflicts were absolutely inevitable. (p. 2)

Mead was skeptical of this position because of obvious empirical variations in adolescent "turmoil," even within Western societies (e.g., German youth in the post-World War I era, and young immigrants to the

United States; Mead, 1928, p. 2). Moreover, there was no *scientific* evidence to support the universal storm-and-stress position. In particular, this position was not based on rigorous experimentation, where potential causes are systematically controlled, and eliminated if nonsignificant (p. 3).

Based on her anthropological training, Mead believed this experimentation could be simulated by systematically comparing cultures in terms of how adolescence was structured and experienced. If dissimilar structures produced different experiences—differing amounts of turmoil, or none at all—then the biological position would be called into question on scientific grounds. Thus, the anthropologist could ask the question: "Are the disturbances which vex adolescents due to the nature of adolescence itself or to the civilization?" (p. 11).

The remainder of Mead's position and experimental strategy is discussed in chapter 2, so need not be stated here. Based on the study she conducted, however, she concluded that because few of her Samoan informants showed signs of behavioral problems brought on by puberty, "adolescence is not necessarily a time of stress and strain, but . . . cultural conditions [can] make it so" (p. 234). In addition, she found that the experience of puberty made no significant characterological difference. In her words:

> The adolescent girl in Samoa differed from her sister who had not reached puberty in one chief respect, that in the older girl certain bodily changes were present which were absent in the younger girl. There were no other great differences to set off the group passing through adolescence from the group which would become adolescent in two years or the group which had become adolescent two years before. (p. 196)

Mead argued that the difficulties affecting the adolescent of the 1920s United States were caused by "conflicting standards and the belief that every individual should make his or her own choices, coupled with a feeling that choice is an important matter" (pp. 234–235). In contrast, she asserted that in 1920s Samoa:

> The growing child is faced by a smaller dilemma than that which confronts the American-born child of European parentage. The gap between parents and children is narrow and painless, showing few of the unfortunate aspects usually present in a period of transition . . . essentially the children are still growing up in a homogeneous community with a uniform set of ideals and aspirations. (p. 273)

With this evidence of dramatic cultural differences in the prevalence of adolescent difficulties, Mead concluded that storm and stress cannot be considered a biological inevitability. She was the first to provide evidence for this conclusion, and as we see here, this conclusion has held even as it has been put to the test with contemporary theory and methodology.

THE CONTROVERY SPARKED BY FREEMAN'S BOOK: MEAD ON TRIAL

Not one to mince words, Freeman (1983) asserted that "Mead's account of Samoan culture and character is fundamentally in error" (p. xii). In diametric opposition to Mead's account, Freeman argued that Samoan culture was/is anything but casual: There was/is a high level of aggression and violence among Samoans; that Samoans idealize(d) premarital chastity, but also commonly engage(d) in rape; and that they were/are highly competitive and easily insulted. In a move that escalated the controversy and kept it alive for the past decade, Freeman placed Mead's study squarely in the middle of the nature–nurture debate. He did this by arguing that Mead was at some level trying to please her mentor, Franz Boas, by giving him conclusive evidence that culture, not biology, determined adolescent behavioral disturbances.

Freeman declared that his book provides "detailed empirical evidence to demonstrate that Mead's account of Samoan culture and character is fundamentally in error" (p. xii). He also said that this evidence "has the specific purpose of scientifically refuting the proposition that Samoa is a negative instance" of adolescent storm and stress because it shows "that the depictions on which Mead based this assertion are, in varying degree, mistaken" (p. xiii). The literature reacting to these and other claims in Freeman's book now comprises hundreds of commentaries, including dozens of journal articles, reviews, rejoinders and letters, a few books (see Caton, 1990), a documentary film (Heimans, 1988), and even the "Phil Donahue Show" (Holmes, 1983a).

The fact that Freeman's arguments often stand in stark contrast to Mead's account is just one indication that an evaluation of the controversy is not a simple task. Furthermore, he has managed to attract a number of followers who have taken up his cause (see especially Appell & Madan, 1988; Caton, 1990). To deal with the polarization of opinion on this matter, it is necessary to consider various claims in light of possible ulterior motives.

Complicating matters further is the fact that life on the remote island Mead studied has changed dramatically. Life there is now rather "Americanized," replete with a high school, television reception, popular Western music, ghetto blasters, pickup trucks, and air service twice a day (cf. Holmes & Holmes, 1992). Because we obviously cannot go back in time to 1925, we have no choice but to rely on inferences derived mainly from written documents and to arrive at judgments of the plausibility of certain evidence and arguments.

In getting to the heart of the matter, therefore, we must stand back and look at the controversy itself, and the characters involved in it. Unfortunately, Mead died in 1978 and did not leave a very extensive defense of her position. In attempting to conceive of what she might say now in her own defense, we run the risk of adopting an overly sympathetic stance toward her work. On the other hand, many reviewers of Freeman's book argue that it is difficult to sympathize with him. This is so in terms of how he originally presented his criticisms and how he has responded to commentary about his criticisms. As Brady (1991) argued, "Freeman has made it difficult to discuss his work without discussing him. . . . There is in many of Freeman's responses a verve, a relish to fight, a sense of retribution far beyond the ordinary. . . . He conveys the distinct impression . . . that to disagree with him is to be morally wrong . . ." (p. 449).

Based on my review of his critique of Mead's work (e.g., Côté, 1992), I am of the opinion that Freeman's intransigence would be more fathomable if he had "hard" evidence supporting tight arguments. As we see in the chapters to follow, however, his evidence often has obvious limitations and his arguments often have alternative conclusions. He did not acknowledge either possibility. Thus, it is hard to sympathize with an unrelenting campaign against Mead's work, which is supposedly simply trying to set the record straight. Incidentally, it is interesting to note that a book critiquing her work was released in the same year, but its author apparently did not believe it necessary to launch a crusade against her to make his point (Goodman, 1983).

The unrelenting nature of Freeman's campaign leaves one with the impression that he is settling a score—but then again perhaps he is. For instance, he openly admitted that he wrote his book partly at the urging of some of his chiefly Samoan friends who were displeased with the image of Samoans that Mead supposedly created (Freeman, 1983, p. xv). As argued here, his disclosures in this regard suggest that he may be playing the role of the Samoan chief who is trying to preserve the honor of Samoans against Mead's perceived negative portrayals. If this is the case, Freeman might have been too close to the culture to have given us the dispassionate analysis he claims to have given.

THE CENTRAL FIGURES IN THE CONTROVERSY

Both of the central figures in this controversy are dynamic and colorful individuals. The debate stimulated by their works undoubtedly reflects their respective characters. It is therefore useful to consider the role played by both figures in this affair. It is also helpful to frame this dispute in terms of a legal metaphor, because it involves charges laid by one person against another. Moreover, the alleged "crime" took place some time ago, and there is little direct or material evidence upon which to assess "guilt" or "innocence." Accordingly, readers are asked to play the role of jury members for this section, and for the remainder of the book. This role requires readers to: (a) assess the plausibility of arguments, (b) determine the admissibility and validity of certain forms of evidence, (c) judge critically the statements of "witnesses," and (d) decide the extent of culpability of the parties involved.

The Prosecutor: Derek Freeman

Many observers of this controversy have pondered over Freeman's behavior in the matter, particularly his persistence and intransigence (e.g., Brady, 1991). Previously, a connection was drawn between the nature–nurture debate and the appeal of Freeman's book for some of his followers. In this section, we look deeper into Freeman's behavior and speculate as to why he undertook his crusade against Mead in the first place; why he has been unwilling to revise any of his criticisms despite obvious evidence to the contrary; and why he persists in trying to prove himself absolutely right and Mead (and others) absolutely wrong.

Freeman, now Professor Emeritus at Australian National University, began his research in Western Samoa in the 1940s when he was in his 20s. During a 2-year stay, he became fluent in the language; was "adopted" by a Samoan *matai* (a titled family head, or chief); and was even granted a symbolic chiefly Samoan title, thereby giving him access to the *fono* (village assemblies of *matai*). This placed him, he claimed, "in an exceptionally favorable position to pursue [his] researches into the realities of Samoan life" (p. xiv).

If we assume that Freeman has identified strongly with the *matai* role and has appropriated that perspective, then we can surmise that he has also internalized the concern with maintaining the ideals of the culture by ensuring that transgressions of the social order are properly dealt with. Primarily, *matai* are invested with the authority and responsibility to direct daily economic and social activities. This includes the assignment of work projects, the distribution of food and available resources, and the meting out of fines and punishments. Secondarily, *matai* are expected to assume an ahedonic, circumspect posture in engaging in the day-to-day life of the community (Holmes, 1987), and to maintain a certain "detachment" from their families, especially from children and young people (cf. Schoeffel & Meleisea, 1983).

In understanding Freeman's motivations in this controversy, then, it is useful to know that his reference group in Samoa is that of the *matai*; much of his knowledge of Samoa is derived from the perspective of its male culture; and he is likely to have been socially and emotionally removed from Samoan young people. Thus, we can understand in part why his portrayal of Samoan society dwells on violence and crime as well as on its authoritarian features. Readers will also recall that Freeman divulged that he had been asked to correct Mead's portrayal of Samoa. In his words, some "educated Samoans . . . entreated [him] . . . to correct [Mead's] mistaken depiction of the Samoan ethos" (p. xv). But, one final piece of information helps to complete the picture. He also disclosed that he knew for several decades that he was going to attack Mead's work. In his words: "By the time I left Samoa in November 1943 I knew I would one day face the responsibility of writing a refutation of Mead's Samoan findings" (p. xiv). Note that he spoke of a *refutation*, not a critical evaluation. In other words, his conclusion had already been formed in 1943. With these statements, Freeman seems to have admitted that long before he wrote his book, he had accepted an assignment as a *matai* to neutralize what was perceived as a blight on Samoan honor perpetrated by Mead.

Samoan author Wendt saw Freeman's motivation in terms of carrying out a "mission." In A. Wendt's (1983) view, Freeman is "a *matai* who takes his mataiship seriously. Most importantly, he has deep love and respect for us. This I think helps explain his almost *obsessive quest to correct what he deems was the wrong Margaret Mead did to us.*" Wendt also argued that Freeman's book is "apt justification for his life, his *mission*, his search to belong. To us " (pp. 12, 69, italics added).

This "mission" would help to explain why Freeman has been determined to disgrace Mead and her associates. At the very least, as mentioned earlier, the degree of his involvement in the *matai* role raises questions about his capacity for detachment and objectivity in the controversy. Indeed, in a documentary film discussed in chapter 2, Freeman is shown weeping during a *fono* meeting (in Western Samoa) in response to the numerous expressions of praise and honor bestowed upon him by his *matai* friends.

Freeman's identification with the *matai* role and his acceptance of this mission also help to account for his intransigence in the controversy. This is especially the case with regard to the issues of sexuality and virginity, and his insistence that social changes between the 1920s and 1960s are irrelevant to the controversy. Modern Samoa is devoutly Christian, and the *matai* and pastors share village power. As part of contemporary Christianized Samoa, the *matai* therefore have a responsibility to ensure that Christian ideals are protected. However, as is seen in chapters 4 and 5, the missionaries encountered a culture they considered "heathen" and "immoral," so they set out to "Christianize" it and to remove all vestiges of this "heathen immorality." Therefore, for Freeman to admit that pre-Christian Samoans practiced any thing other than Christian sexual behavior would be to undermine the current *matai*-pastor power sharing and to violate the mandate of the *matai* role, because it would validate this "heathen" past.

To continue this line of reasoning, it would appear that Freeman "the scientist" cannot admit that Samoan sexual practices were significantly different before Christianity, because Freeman "the *matai*" cannot admit that Christianity has had any significant impact on elements of Samoan society related to sexual and social mores. Thus, it appears that Freeman has had to draw a line of defense on these points, because any other position would compromise his stature as a titled Samoan, even though it is a symbolic title. Incidentally, it is of interest to note that O'Meara (1990) observed, in the village he recently studied, that most people "describe their own, pre-Christian past as the time before they 'saw the light.' They rarely speak of that time, and then only with some embarrassment" (p. 47).

Brady (1991) drew a similar conclusion, linking Freeman's motivation regarding this mission on behalf of Samoans with his position on the nature–nurture debate. He noted as well that some of Freeman's supporters recognize this: "that Freeman has been on a heroic and highly personalized mission—a 'double-edged' quest—to discredit Mead, first on behalf of what he declares is superior science and second on behalf of some very conservative Samoans who have found Mead's characterizations of their attitudes about sex embarrassing and offensive" (p. 499).

To put Freeman's strategy in a context more familiar to Westerners, his tactics resemble those of an aggressive prosecuting attorney who does what is necessary to win a "case." For example, in his book, he based some of his critique on the premise that Mead was young (age 23), inexperienced, and apparently anxious to please her mentor, "Papa Franz" (Boas). This, he said, led her to draw some hasty conclusions in order to provide Boas with much needed evidence for his position in the nature–nurture debate. Such *ad hominem* aspersions are irrelevant and unprofessional, and have contributed to a degeneration of this "scientific debate." This *ad hominem* reasoning could just as easily be turned back on Freeman with similar irrelevant charges. For example, although I would not do so, someone could just as easily cite him as an eccentric old man trying to live up to Karl Popper's reputation, the person to whom he dedicates the book. But, such name calling accomplishes nothing for science.

More recently, Freeman (1991) chronicled what he claimed to be the history of Mead's involvement before, during, and after her research in

Samoa. In a serious departure from academic protocol, Freeman's "historical analysis of the Samoan researches of Margaret Mead" can be described as little more than "soap opera science." Although this judgment may seem harsh, readers will find in that article rumor and innuendo, hearsay and gossip, and unsubstantiated insinuations about Mead's character. Is it really germane to a scientific dispute whether or not Mead had a one night stand at a conference, whether she and Ruth Benedict had "an intimate Sapphic relationship" (p. 105), or whether she danced barebreasted when in Samoa? I think not. In fact, there is no solid scientific evidence in that article that can be said to resolve anything, other than prurient and voyeuristic desires.

Others have objected to Freeman's patronizing attitude toward Mead. Turnbull (1983), for example, argued that "[w]ith a revealing show of male chauvinism, Freeman even mentions how this 'young female student' included as part of her field equipment some 'cotton dresses.' This irrelevant piece of information he gleaned from his avid reading of Mead's writings, scholarly, popular, and private, randomly mixed to cast her not only as a mindless Boasian, but as a rather empty-headed, silly girl" (p. 33).

As we see, Freeman's evidence against Mead is circumstantial. Although he postulated a motive (he argued that she was attempting to please her mentor, Franz Boas) and opportunity (her selection of a research site that she thought in advance would likely validate her hypothesis), he did not have any material or direct evidence from Ta'u in the 1920s pertaining to most of the matters involved (except for the flawed testimony of one person, described in the next chapter). Therefore, he did the two things that any ambitious prosecutor would do: (a) he provided as much circumstantial evidence as can be mustered; and (b) he raised questions about the character, morality, and competence of the accused. With a skillful presentation, he attempted to sway the jury of public opinion to convict Mead of his charges. As is seen in this volume, however, when a reasonable defense is provided for Mead, his evidence is far from convincing or conclusive.

The Defendant: Margaret Mead

The second principal figure in the controversy is Mead herself. In this section we see how her own behavior appears to have contributed to both the current bad press about her book and the loss of reputation she has suffered personally. As we see, the extent to which she wrote *Coming of Age* as a popular book has contributed to this controversy, as has the way she handled criticisms of that book since it was published. We also see, however, that these failings do not justify the serious indictments raised by Freeman.

As an academic, Mead could have handled matters much better, beginning with the initial publication of her book in 1928. This is the wisdom of hindsight, however, and for the young academic the prospect of publishing can make other concerns seem unimportant. At the time, she was new to the scene and this book was to lead to immediate recognition and acclaim. To make the book marketable, her publisher apparently urged her to write additional chapters in a popular style, and it is primarily the content of these chapters that has been the object of criticism. Certainly, these chapters contain some generalizations that are difficult to defend academically (see chapters 3 and 6, this volume).

To put the issue of marketability into perspective, Mead's (1930/1969) scholarly report of her Samoan research, *Social Organization of Manu'a* would never have been published in place of *Coming of Age* by a popular press. If she had wished to stay academically "pure," however, she should have only published something like *Social Organization*. In any event, academics who are dissatisfied with the offending chapters from *Coming of Age* should read, for example, the following segments from *Social Organization*: "Daily Life in Manu'a" as a substitute for "A Day in Samoa", and "Dominant Cultural Attitudes" as a substitute for "Our Educational Problems in the Light of Samoan Contrasts."

It also appears that Mead did not understand Samoan culture as well as she thought (Holmes, 1987; Schoeffel & Meleisea, 1983). Although she did spend 9 months intensively and directly studying the culture, even after years of study, elements of any culture may not be fully appreciated by an outsider. In particular, she may not have fully understood the "casualness" of Samoans, but she would not be the only one to have had this problem (Swaney, 1990). Clearly, Westerners accustomed to a time-managed, hurried, compartmentalized existence have a difficult time understanding the "laid-back" nature of some non-Western cultures. It is common that apparently "relaxed" behaviors and attitudes are mistaken by Westerners for laziness or dullness, or in Mead's case, an emotional shallowness. From the accounts of Samoans commenting on the controversy, however, Freeman may not be any closer to understanding the "true" Samoan character, if any such thing exists (cf. Ala'ailima, 1984; Schoeffel & Meleisea, 1983; A. Wendt, 1983; F. Wendt, 1984).

Eventually, Mead did entertain the notion that her informants provided her with a skewed view of Samoan culture. Unfortunately, she should have considered this problem much sooner, and she should have at least warned readers in the preface of later printings. After all, much of her information came from young women who were "coming of age" and by Mead's own account were experiencing the best years of their lives. Her informants were temporarily free from many responsibilities and obligations, and most were able to experiment socially in a playful fashion.

One's culture can look very benign from this point of view. Such a "moratorium period" is by definition one where individuals are largely free from responsibilities and irreversible commitments (Mageo, 1988; cf. E. Erikson, 1968). During a moratorium the "stakes" are often not "high" for those involved, because they are relatively free to casually move from one interest to the next.

The possibility that Mead's perspective was somewhat askew does not mean, however, that Freeman's was any more accurate. As we saw in the last section, Freeman's appropriation of the *matai* role and his mission to counteract Mead's supposed influence may have produced a narrow lens through which he viewed both Samoan culture and scholarly criticisms of his work.

Finally, although it is possible to give Mead the benefit of the doubt in most instances, she was undoubtedly in error in stating that "because one girl's life was so much like another's, in an uncomplex uniform culture like Samoa, *I feel justified in generalizing* although I studied only fifty girls in three small neighboring villages" (Mead, 1928, p. 11, italics added).

Not only is it unclear to whom she is generalizing, it is also unclear as to where and when. She wrote that she went to Ta'u "as a corrective for the degree of culture contact" elsewhere in American Samoa (Mead, 1930/1969, p. xv). Thus, there is an obvious flaw in her logic if she selected the island of Ta'u because of its remoteness, but still felt justified in generalizing to the rest of Samoa. Why bother going to Ta'u when by the logic of the quotation just cited she could have found the same thing anywhere in Samoa? Logically, she cannot have it both ways. Moreover, she was also aware that what she observed on Ta'u was no longer completely "traditional" either, and that the culture there was undergoing changes when she conducted her study. Given the uniqueness of Ta'u, then, there are grounds to question some of Mead's generalizations to the rest of Samoa. But, by the same token, Freeman's contention that Ta'u in 1925 would have been no different from Western Samoa in the 1960s is questionable (we explore this further in chapter 3). In Appendix III of *Coming of Age*, Mead discusses the historical uniqueness of the "Samoa" she studied on Ta'u.

As for the "real" Samoa, I do not think that anyone, even Samoans themselves, can lay claim to a complete understanding of Samoan culture in all of its forms in all places and times. If anything, Freeman may have simplified it as much as Mead is thought to have (Ala'ailima, 1984; Bock, 1983; Feinberg, 1988; Hooper, 1984; Levy, 1984; A. Wendt, 1983; F. Wendt, 1984). Moreover, the resolution to the controversy is not likely to be found "somewhere in between" the two versions, as some have suggested (e.g., Swaney, 1990). Rather, it appears to lie in part with a complex and flexible culture that survives and maintains its integrity by appropriating the forces that seek to change it, and by defying easy definition and understanding (cf. Laing, 1987). The protective strategy of incorporating an influence as much as possible before the influence incorporates the culture was noted by Mead, and has since been noted by others (e.g., Baldauf & Ayabe, 1977). This is believed to be the case with Christianity, for example. A number of observers have remarked that it is difficult to say whether Samoa was Christianized, or whether Christianity there was "Samoanized" (e.g., Hanson, 1973; Holmes, 1980a, 1980b; Meleisea, 1987a; Stanner, 1953).

In any event, it is important to note that Samoan culture is complex and is based on a long history of intricate and subtle customs and oral traditions. It currently has a reputation as the most conservative of the Polynesian cultures (Hanson, 1973; Holmes, 1980a), and Samoans proudly refer to their way of life as *fa'a Samoa* (the Samoan way).

In light of these considerations, had Mead initially titled the book something like *Coming of Age on Ta'u*, and not engaged in wholesale generalizations to other Samoan islands and to other historical periods, there would not likely be a controversy today about her work. Indeed, in a later scholarly publication, she exercised more caution in stating that her generalizations regarding Samoan should be "understood to refer to Manu'a specifically" (Mead, 1931, p. 545). Nevertheless, there is a controversy. In the next chapter we begin the task of judging the merits of the charges made by Freeman that stimulated one of the greatest disputes in the history of anthropology.

METHODOLOGIES APPROPRIATE TO THE EVALUATION OF MEAD'S COMING-OF-AGE THESIS

Given the extreme and disparate positions in this case, how do we go about evaluating Freeman's charges, as well as Mead's original work? Before proceeding, it is appropriate to apprise readers of the methodologies I employed to evaluate the controversy.

As mentioned earlier, because life has changed dramatically at Mead's research site, there is little point in attempting to replicate her study there. Instead, we are confronted with the task of piecing together available evidence until enough of an understanding is developed to allow for a conclusion to be drawn. What I have done, therefore, is to undertake an interdisciplinary analysis, utilizing a number of different qualitative and quantitative methodologies.

In chapter 2, where Freeman's case against Mead is laid out, I "deconstruct" the politics of the controversy, to discover that it is in many ways yet another battle in what is for many people a sort of "nature–nurture war." Indeed, when we systematically examine the components of the arguments and evidence that have been assembled in the case against Mead's coming-of-age thesis, we find that a number of interesting maneuvers have been undertaken. One such maneuver is a documentary film that claims to resolve the controversy. When this film is deconstructed, however, and the *scientific* merit of the evidence presented in this film is examined, a different conclusion can be drawn about what this film means for the Mead–Freeman controversy.

In chapter 2, I "unpack" the statistics Freeman produced to attempt to refute Mead's coming-of age thesis. Freeman presented a number of statistical arguments concerning crime and delinquency that he believes exposes Mead's coming-of-age thesis as erroneous. To evaluate his critique, these statistical arguments are critically examined with state-of-the-art empirical knowledge regarding adolescence and human development. This critical evaluation is carried out on the data that were collected by Freeman in Samoa, as well as on comparative data he used from other countries. Similarly, the crude crime rates Freeman used to argue that Mead's thesis was wrong, are algebraically converted into estimates of what life would be like on a daily basis in a given Samoan village. Thus, Freeman's statistical case against Mead is examined in terms of statistical realism and errors of statistical conclusion.

In chapter 3, several forms of evidence are examined, including expert opinion and material evidence I gathered during a fact-finding visit to Western Samoa and Mead's research site in American Samoa. This material evidence, which includes geographic, topographic, and photographic material, provides us with reference points with which to evaluate Freeman's claims.

Chapters 4 and 5 undertake a sociohistorical analysis of adolescence in Samoa since contact with the West. Based on textual analyses of the detailed diaries kept by Christian missionaries and the ethnographies of early anthropologists, a trajectory of social change is mapped out, giving us a working knowledge of adolescence in Samoan culture over the past 150 years. Chapter 6 presents a summary and analysis of "Mead's Samoa" of the 1920s. With the working knowledge gained in chapters 4 and 5, we are then in a position to judge whether "Mead's Samoa" reasonably fits with this historical trajectory. In this chapter, I also analyze data collected by Mead that were never formally crosstabulated, apparently because Franz Boas told her not to bother. When these data are carefully examined, conclusions can be drawn regarding the normative structure of the villages she studied.

The remaining two chapters of this book deal with the situation that currently obtains in Samoa. Statistics and textual accounts are presented in these chapters that reveal what has happened to how young Samoans come of age since Mead conducted her study in the 1920s. This empirically derived understanding of contemporary Samoa provides a basis for policy recommendations, as well as recommendations for future research, with respect to the problems confronting those currently struggling to come of age in a rapidly changing Samoa.

CHAPTER TWO

Freeman's Case Against Mead

THE PREMISE OF FREEMAN'S CASE: ADOLESCENT BIOLOGY

Freeman (1983) believes he has proven that Mead "was at error in her depiction of the nature of adolescence in Samoa." Based on this belief he declares that "her much bruited 'negative instance' [regarding adolescent "storm and stress"] is . . . no negative instance at all" (p. 268). In other words, Freeman argued that Mead did not find a culture where adolescence was free of storm and stress, so he believes she did not have evidence that refuted the biological argument of adolescent turmoil. He did not waver from this position, and more recently referred to Mead's coming-of-age thesis as "preposterous" (1985, p. 910) and "egregious" (1987a, p. 930) with its conclusion that adolescent storm and stress is not primarily biological in origin.

When the basis of this strong condemnation is examined, however, it appears that Freeman misconstrued Mead's conclusion. From his reconstructions of Mead's writings, Freeman appears to be attempting to portray her as professing that "biology" has no relevance for adolescent behavior or disturbances in behavior during adolescence. In one place he "quoted" her as saying that " 'we cannot' . . . 'make any explanations' in terms of the biological process of adolescence itself" (1985, p. 910); in another he contended she said that " 'we cannot make any explanation' of the 'disturbances' of adolescence other than 'in terms of' the 'social environment' " (Freeman, 1987a, p. 930).

The reader will note that these are not direct quotes; rather, they are collages of passages that, as we see later, do not accurately represent Mead's conclusion. Freeman has these "creative" collages throughout his book. To make this matter more problematic, he provided all of the references from each paragraph in single endnotes, so that it appears in his text that the quoted passages are from one source. This technique produced more than 50 pages of endnotes. Moreover, in many of these collages, Freeman is not quoting from Mead's *Coming of Age*, but from other sources, published for vastly different audiences over her 50-year career. This unusual referencing style makes it difficult for the reader to judge the validity of Freeman's reconstructions of Mead's thoughts. As Rappaport (1987a) noted, "Professor Freeman seems to have combed Mead's corpus for phrases and sentences that can be construed in ways convicting her of his charges" (p. 304). See McDowell (1984) for a thorough critique of Freeman's writing and referencing style.

As we saw in chapter 1, Mead's conclusion is actually quite different. She simply said that because the biological processes associated with adolescence are the same for Samoans and Americans, many difficulties experienced by U.S. adolescents may be attributed to characteristics of U.S. culture that are absent in Samoan culture. This reasonable conclusion was stated as follows by Mead (1928):

> If it is proved that adolescence is not necessarily a specially difficult period in a girl's life—and proved it is if I can find any society in which that is so—then what accounts for the presence of stress and storm in American adolescents? First, I may say quite simply that there must be something in the two civilizations to account for the difference. *If the same process takes a different form in the two different environments, I cannot make any explanations in terms of the process, for it is the same in both cases.* But the social environment is very different and it is to it that I must look for an explanation. What is there in Samoa which is absent in America, what is there in America which is absent in Samoa, which will account for the difference? (pp. 197–198, italics added)

A fair reading of this full quotation suggests that Mead is simply saying that the biological processes associated with adolescence are the same in both cultures, not irrelevant. In other words, she identified the biology of puberty as a *constant* and social environment of adolescence as a *variable*. Mead (1928) provided the following justification for taking this position:

> For the biologist who doubts an old hypothesis or wishes to test out a new one, there is the biological laboratory.... Keeping all the conditions but one constant, [the biologist] can make accurate measurement of the effect of one. This is the ideal method of science, the method of controlled experi-

ment, through which all hypotheses may be submitted to a strict objective test. (p. 5)

Thus, Mead was aware of the requirements of the experimental method and of biological research in this area. Her design was sound unless one is willing to argue that pubertal biology is linked with "racial" differences. That is, a potential flaw in the logic of her design is that the variable of "race" is not held constant in her comparison between U.S. and Samoan cultures. Although some of Freeman's supporters have suggested that this may be a problem (Appell, 1984; Buchholz, 1984), they should heed Sprinthall and Collins (1984) who concluded that "racial and ethnic differences in puberty are minor" (p. 74). In any event, Freeman has not taken her to task on this matter.

Interestingly, in his original critique, Freeman did not paraphrase Mead's conclusion; instead, he provided it verbatim. After doing so, he stated the following:

In other words, any explanation in biological terms of the presence of storm and stress in American adolescence was totally excluded. . . . Instead of arriving at an estimate of the relative strength of biological puberty and cultural pattern . . . Mead dismissed biology, or nature, as being of no significance whatsoever in accounting for the presence of storm and stress in American adolescents, and claimed the determinism of culture, or nurture, to be absolute. (Freeman, 1983, p. 78)

Freeman's interpretation of Mead's conclusion raises questions about his understanding of both research design and biological research on adolescents. As just argued, Mead had, in principle, a sound research design. More specifically, U.S. society constituted the control group and Samoan society the experimental group; the independent variable was cultural institutions governing adolescence (known to differ) and the dependent variable was the ease/difficulty of coming of age. If a difference were found on the dependent variable, it would be attributable to cultural institutions, because the biological processes of puberty were essentially held constant.

In spite of this reasonable research strategy, Freeman (1983) contended that Mead was not "scientifically equipped to investigate the subtle and complex interaction, in Samoan behavior, of biological and cultural variables" (p. 75). These criticisms are unconstructive because of their vagueness, however. For example, he never specified just what she should have done to assess biological variables. In 1925, the state of knowledge regarding the biology of adolescence was totally theoretical and there was virtually no technology available to measure relevant physiological variables.[1] In fact, little scientific attention has been paid to the relationship between puberty and adolescent behavior until quite recently. As noted by Brooks-Gunn and Reiter (1990):

> until the 1980s almost no developmental work on the meaning or effects of puberty had been conducted.... Then in the 1980s, starting with a 1981 conference on girls at puberty, this topic came into its own. During that decade at least six conferences were held on early adolescence, with the topic of puberty having a prominent place in each. (p. 19)

Freeman also claimed that Mead did not carry "out any systematic comparison of hereditary and environmental conditions" (p. 76). To the contrary, with puberty defined as the onset of menstruation (Mead, 1928, pp. 283–284), Mead assembled a sample to represent prepuberty ($n = 11$), imminent puberty ($n = 14$), and postpuberty ($n = 25$) and she noted that there were no major characterological or psychopathological differences among these three groups. If puberty is the "hereditarian condition" responsible for adolescent storm and stress, such differences should have been observed among such a grouping of informants. This research strategy is apparently inadequate for Freeman, however. Again, just what he expected her to do is unclear, as is what he would expect of someone conducting such research today with current knowledge and technology.

In view of these considerations, Freeman's critique of Mead's treatment of the biological factors affecting adolescence is questionable. If we assume that he is not being obtuse in his reading of Mead, we might conclude that he is naive regarding the difficulties of understanding and documenting the interplay between biological and cultural factors. Moreover, at no time in her career did Mead dismiss the relevance of biological factors (cf. Rappaport, 1987a, 1987b). If she were able to participate in this debate today, she probably would clarify her position by arguing that cultural factors can stimulate certain biological reactions in adolescents that are part of a chain of causation in some of the storm and stress associated with this period. She was simply implying that in the absence of certain cultural factors, these biological reactions are not stimulated and that therefore the storm and stress can be considered neither natural nor inevitable.[2] She would therefore probably not have

---

[1] In fact, Mead did look for some sort of technology to help her, but all that was available at the time was the newly developed instrument to measure galvanic skin response. She concluded that it would not have been suitable for the task, however, partly because it would only measure gross emotional responses (Mead, 1972).

[2] See Paikoff and Brooks-Gunn (1991) for a discussion of research models that need to be tested to sort out such issues.

argued that biological factors are "of no significance in accounting for the presence of storm and stress in American adolescents" (Freeman, 1983, p. 78), but that they constitute intervening rather than independent variables. As Rappaport (1987a) noted:

> Mead clearly recognized the biological character of puberty, never claimed that biological factors had nothing to do with behavior, and simply stated that differences in the emotional and cognitive correlates of "the same [biological] process" in "different environments" are to be accounted for by environmental differences. *Hardly preposterous.* (p. 160, italics added)

Ironically, recent research undertaken with state-of-the-art theory and technology supports Mead's conclusion that biological factors do not directly affect most adolescent behavior or disturbances (e.g, Adams, Montemayor, & Gullota, 1990; Montemayor, Adams, & Gullota, 1990). Referring to the notion that hormones are primarily responsible for emotional turmoil among adolescents, Brooks-Gunn and Warren (1989) noted that in the empirical literature "hormone-affect associations, when found, . . . are small. . . . No direct effects of pubertal status were found" (p. 47). In their own study of the negative affect of 100 females, age 10–14, Brooks-Gunn and Warren (1989) found that "endocrine system factors . . . accounted for no more than 4% of variance in negative emotional expression . . . [while] social factors accounted for more variance (8% and 18% for depressive and aggressive affect, respectively) than biological factors" (p. 47). They concluded from these findings that "hormonal activation effects in humans may not be large . . . and may be overshadowed by environmental events."

Other experts in this area agree. For example, after reviewing the relevant literature, Montemayor and Flannery (1990) concluded that the evidence indicates "that pubertal effects, by themselves, account for a relatively small proportion of the variance in early adolescent behavior" (p. 294). Sprinthall and Collins (1984) provide this elaboration of the implications of state-of-the-art research:

> biological changes [associated with puberty] seem to influence psychological development through the subjective meanings that the changes have for adolescents themselves and for adults and peers around them. . . . [Thus] the effects of the primary physical changes of adolescence are *socially mediated* by the reactions of self and others. In this view one's self-image and self-esteem reflect one's own and others' subjective reactions to biological maturation. And these reactions are determined by sociocultural standards, norms, and expectations about physical characteristics that are widely held in a society or culture. (p. 79)

Finally, in specific reference to the Mead–Freeman controversy, Muuss (1988) maintained that:

> [Freeman's] argument for a synthesis between biological and cultural determinism seems to come too late to have a major impact. The synthesis has already taken place. Although proponents of culture or biology may emphasize one more than the other, most contemporary theorists view development as an interaction between nature and nurture. *Adolescent turmoil, storm and stress, and crises are no longer considered inevitable, not even for adolescents growing up in the United States.* And regardless of whether or not Mead was correct about Samoa, other anthropologists have observed societies in which adolescence is not turbulent. Thus, the physiological changes of puberty and sexual maturation alone are not categorically responsible for adolescent difficulties. Anxieties, insecurities, social pressures, social expectations, and cultural, educational, and family factors may all contribute to adolescent stress. (p. 142, italics added)

In summary, when critically examined, Freeman's critique of Mead's treatment of biological factors amount to straw arguments. As Rappaport (1987a) noted: "The conclusion that Freeman asserts was Mead's [conclusion] is indeed, preposterous, *but it verges on the preposterous to attribute [that conclusion] to her*" (p. 304, italics added).

Why did Freeman go to such great lengths to construct this argument? The answer appears to lie with his position on the nature–nurture debate and the perception that Mead's study constitutes a cornerstone of the nurture side of that debate. If we assume that he had already set out on the mission to redeem the "honor" of Samoans (as argued in chapter 1), the inclusion of this argument gave him the opportunity to attempt to score a victory against those whom he thinks have given insufficient attention to the biology of human development. In doing so, he was apt to gain allies from those who favor nature explanations of human behavior. We see in the next section how Freeman gained allies by attacking Mead's coming-of-age thesis, thereby creating the perception that he has delivered a blow to the nurture side of the nature–nurture debate.

THE POLITICS OF THE CONTROVERSY

The politicking behind the controversy is an interesting sociological phenomenon on its own. As such, it deserves detailed consideration here, because it helps us understand why a small study done more than 60 years ago commands so much attention today. In this controversy, the world has witnessed anthropologist pitted against anthropologist, publicly insulting and discrediting each other. Throughout the brouhaha that ensued from the highly publicized release of Freeman's book in 1983, the central focus of Mead's study has often been ignored; instead, her book became symbolic of much larger fractures within the anthropological community—in fact, within the social scientific community. As we see, when viewed in terms of possible ulterior motives, the "real reason" for much of the attention given to the Mead–Freeman controversy appears to be that it is yet another battle in the nature–nurture war, if you will, that began over a century ago.

Freeman prefaced his attack on Mead's research with a four-chapter treatment of what he believes to be somewhat of a cultural determinist conspiracy that has fabricated evidence for the nurture side of the debate. To counteract this conspiracy, he seemed to think it is important to refute Mead's study because it provides a "negative incidence" that discredited the nature position of absolute biological causation regarding adolescent storm and stress. Readers of his book must wade through this material before they get to the substance of his critique of Mead's actual research. Yet curiously, Freeman eventually acknowledged after some 255 pages that "research on ordinary adolescents has generally failed to substantiate claims of the inevitability and universality of adolescent stress" (Freeman citing Katchadourian, 1977). In other words, Freeman admitted the validity of Mead's basic coming-of-age conclusion!

Why, then, did he write his book? To answer this question we could fall back on the explanation that he is really trying to settle a score for his chiefly Samoan friends, as mentioned earlier. Although this does appear to account for some of his motivation, had his book simply focused on the dissatisfaction of some Samoans with Mead's portrayal of Samoan culture, I doubt that it would have been published in the first place, or that it would have drawn so much criticism and support. In short, there seems to be something more potent to this controversy—something that has drawn otherwise circumspect social scientists into a rather unpleasant row. Thus, one is left with a strong suspicion that it is the nature–nurture debate that provides much of the potency for the controversy.

Given his attack on the nurture side of the debate, one would predict that Freeman's followers in the controversy would be nature advocates. In fact, Freeman's "exposé" of Mead can be seen as a counterstrike to the bad press the nature side of the debate has received over the years. This bad press includes recent revelations that the British psychologist Sir Cyril Burt published fraudulent results during his lifelong quest to prove that intelligence is for the most part genetically inherited (see Broad & Wade, 1982; Gould, 1981). What Freeman's work promised was that finally a "victory" could be scored against the nurture side, and for once it would not be the nature side being accused of deceit.

This possible ulterior motive in Freeman's "refutation" of Mead would account for why he painted such a dark portrait of Samoan culture and character in spite of the risk that it would displease Samoans who object to such harsh portrayals of them and their culture. That is, to appease nature advocates he may have been trying to prove that traits such as aggressiveness are universal human "biological imperatives" that cannot be mitigated by culture, contrary to what Mead's study implied.

The appeal of Freeman's attack on Mead can be seen in some of the commentary provided by his followers. For example, shortly after the controversy began, Caton (1984) claimed that in working through the "Mead Problem" (p. 29), anthropology will likely "undergo a complete reorientation" (p. 29). By exposing Mead and returning "Samoa to reality," Caton submitted that anthropology will be reformed. His logic is stated as follows:

> Those who accept Freeman's refutation affirm a set of observations that exhibit cross-cultural constants operating throughout Samoan culture. The refutation consequently entails rejection of cultural determinism, which denies those constants. One is obliged to accept a model that predicts and explains the constants. That model, Freeman amply documents, is the interactionist model of nature and nurture, based upon behavioral biology.
>
> The other option is to defend Mead's fantasy. That defence must take the way of disingenuous evasions, and pay the price of bad conscience; or else it must become intransigent obscurantism fighting with epithets and dirty tricks. In America especially obscurantism has been highly visible. (pp. 30–31)

These are strong words, but Caton continued railing against the U.S. anthropological "establishment" for its reaction to Freeman's book:

> We see here operating the obscurantist mind that equates behavioral biology with conservatism on the ground that acknowledging significant biological control of behavior is incompatible with social improvement. *This is the gut issue. The Americans are paranoid about biology*, and indeed about any facts inconsistent with the wish to improve the species out of sight. *What's the matter with them? Why are they so discontent with the species as it is?* (p. 31, italics added)

There certainly is much that can be unpacked from this tirade. But, aside from the fact that he did not specify exactly what it is from "behavioral biology" or Freeman's interactionist model that we should wholeheartedly embrace, one has to wonder how he can be so smug about a species that is potentially so destructive, particularly when that species clearly exhibits cross-cultural variation in its tendency toward destructiveness. One does not have to be a social scientist to be aware of the ameliorating effect of particular cultural arrangements on the darker side of "human nature" or on the *potentials* of human behavior. Undoubtedly, few in the U.S. anthropological "establishment" would disagree that the darker side of human nature emerges when stimulated by particular circumstances, and that "human potential" can be stimulated by other circumstances.

Caton (1990) more recently published *The Samoa Reader: Anthropologists Take Stock*, an anthology of articles and news clippings related to the Mead–Freeman controversy. He claimed this book presents a "range of opinion" (p. 1) on the subject. As one reads *The Samoan Reader*, however, it appears that selective evidence is provided in favor of Freeman's critique. This is not surprising because some of the material was provided by Freeman from his personal library (Caton, 1990, p. v). Moreover, Caton, the editor, wrote a section clearly intended to discredit Mead and her supporters (entitled "Did Margaret Mead Speak Samoan?", pp. 158–164. See Brady, 1991, for a review of this book).

Other statements and actions in defense of Freeman have been similarly problematic. For example, Appell and Madan (1988) provided what appears to constitute a position statement for some of Freeman's followers:

> Many of us who view anthropology as a scientific and scholarly discipline have been shaken to our intellectual foundations by the response Freeman's book has received in American anthropology. The attempts to silence him, to deprive him of the right of adequate reply to published criticism, the failure to weigh the issues critically and put them to Popperian test, the conscious and unconscious misreading of his argument including denial of the facts on which it is based, and the attempts to divert attention from his refutation by *ad hominem* attacks have astonished many of us. We find these reactions counterproductive in a discipline that is supposedly seeking the truth. For all those skilled in social analysis and willing to turn this analysis upon their own social lives, *such reactions are the reactions of a belief system that is out of touch with the real world and unresponsive to change*. (p. 23, italics added)

In other words, Appell and Madan seem to believe not only that criticisms of Freeman are unfair, but that criticisms are the result of anthropologists being afflicted with some sort of pathological dogmatism (cf. Rokeach, 1960). Freeman himself is apparently preparing his own sociological analysis of the controversy based on Leon Festinger's sociopsychological classic *When Prophecy Fails* (Freeman, 1987b). He contended that Mead's supporters are exhibiting what he called "the Festinger reaction" because they still support her work "despite the presence of unequivocal and undeniable evidence" to the contrary (pp. 392–393). More recently, Freeman (1991) claimed that Mead was "cognitively deluded" during her study and *Coming of Age* spread her delusion, producing one of "the more spectacular and instructive instances of collective cognitive delusion in the history of the human sciences" (pp. 118–119).

In example after example, the actions of Freeman and some of his followers come across more as a take-no-prisoners attack on Mead and cultural anthropology, than as a dispassionate analysis of the general merits and limits of her work. For instance, among Freeman's followers there have been attempts to support Freeman's blanket claim that Mead was simply "wrong" and that there is little merit in her book. In an early commentary, Appel (1984) said that he "found Freeman's argument to be completely convincing" (p. 189) and several pages later proclaims that "anyone who can read with a discerning mind would have seen that Mead's *Coming of Age* was just plain rubbish" (p. 205).

Appell, an ex-student of Freeman, has since co-edited a book entitled *Choice and Morality in Anthropological Perspective: Essays in Honour of Derek Freeman* (Appell & Madan, 1988). In the first chapter of that book the editors attempt to explain some of Freeman's bullying behavior while at the same time admonishing Freeman's critics for being unfair to him. In their defense of Freeman they noted that although he is sometimes perceived to be "dogmatic and self-assertive," these traits should really be seen "as dedication, enthusiasm, and an unrelenting pursuit of the truth" (p. viii). They also argued that part of Freeman's character is a "deep sense of moral responsibility to espouse the cause of those, whether individual or groups, whom he believes to be the victims of caprice, prejudice, or persecution" (p. ix). As for Freeman's response to criticisms of his attack on Mead, they claimed that he has responded in a dignified manner to "attacks on his work and personal integrity" by answering "each point of criticism in the serious manner befitting a scholar" and that he has "never returned abuse for abuse" (p. ix).

Yet another reviewer announced that Freeman's book constitutes "our most up-to-date and reliable account of Samoan culture" (Badock, 1983, p. 606). This is a most curious statement because Freeman's book is not an ethnography, as he clearly pointed out (Freeman, 1983, p. xii). Such commentaries are unproductive because it is obvious that many of their authors have little or no direct knowledge of Samoan culture or history, or of adolescent development, upon which to assess the plausibility of Freeman's claims. Instead, as suggested earlier, their support for Freeman appears to originate with his attack on the nurture side of the debate and their common complaint that the nature side has not been sufficiently acknowledged by the U.S. anthropological establishment.

It appears that support for Freeman gained enough momentum to spawn a documentary film to support his case against Mead (Heimans,

1988). The film, entitled *Margaret Mead and Samoa*, begins with the narrator claiming that the controversy "will be resolved by startling new evidence presented in this program" (see Freeman, 1989a, 1991, 1992, for his discussions of this "startling new evidence"). To the uninformed, uncritical viewer, the film would appear to present an objective account of the controversy. For example, interview segments are presented with those who seem to represent both sides of the debate. However, an informed, critical examination reveals something quite different. A critical examination uncovers a rather slick production that carefully orchestrates what information is presented, and the pace of its presentation. Moreover, disclosures about the controversy are selective and far more interview segments are presented in support of Freeman's position in the controversy. The few interview segments with those who support Mead's position tend to present only ambiguous or potentially discrediting information about her.

The first portion of the film provides a protracted *ad hominem* portrayal of Mead, referring to her presumed personality characteristics, weaknesses, and frailties as a woman. She is said, for example, to have overcompensated because of her plain looks, to have searched for a mentor, and to have had a terrific need to be accepted by others. Essentially, a negative impression of Mead the person—the woman—is created. This is done presumably to create suspicions about her academic integrity and abilities. Nowhere in the film is there an analysis of Freeman's personality, or an implication that his gender somehow compromises his intellectual integrity and ability. In addition, Ruth Benedict, Mead's close friend and colleague, is described as an unhappily married woman who had done a number of things but had never found fulfillment. The tendency to disparage women who step out of traditional roles may be widespread, but it certainly cannot be excused because of its prevalence, especially in what is supposedly a scientific debate.

With Mead's credibility undermined, the film pretends to represent her Samoan research. However, it focuses almost entirely on the issue of female Samoan premarital sexual behavior, rather than Mead's central thesis—the issue of whether adolescence was problematic for her informants. In presenting this slanted view, a number of points are made that seriously misrepresent both Samoan culture and her research there. Two examples of this are cited here.

In one interview, we witness a curious leap of logic based on the reasoning that because Mead did not report any pregnancies in her book, there must not have been any sexual activity taking place. Not only is such a statement naive, it is ill-informed because it ignores the ample evidence of premarital and extramarital sexual behavior in precontact and contemporary Samoan culture, as well as the continuing practice of the casual adoption of children from unwed mothers (Schoeffel & Meleisea, 1983). The fact is that Mead (1928) cited several examples of "illegitimacy" in her book (pp. 54–55, 56, 153, 168, 181), as well as the technique used for inducing a miscarriage (p. 153).

In another excerpt in the film, someone identifies the double standard in contemporary Samoan society by which males attempt to have sex with as many women as possible, while at the same time protecting their sisters from the sexual advances of other males. The conclusion is drawn for us, however, that females are not having sex before marriage. Although this double standard undoubtedly now exists, as it does in many societies, logic dictates that if some males are having sex, some females must also be having it (the issue is heterosexual activity). In his haste to support Freeman, this "expert" is ignoring Freeman's (1983) own data that reveal that 27% of a sample he drew of unmarried Samoan female adolescents were nonvirgins. Those unfamiliar with the details of this controversy are truly misled by such arguments.

By dwelling on the issue of sexual behavior, and presenting witness after witness who declares that Samoans do not engage in premarital sex, the film builds to a "climax" where the controversy is supposedly resolved by what Freeman (1989a) called "crucially important new evidence." When this evidence is critically analyzed, however, it is obvious that scientifically it is neither reliable nor valid.

With respect to its reliability, it is based on the testimony of a witness of doubtful credibility. The informant, Fa'apua'a Fa'amu, declares that although she and her agemates treated Mead like a *sister*, they also lied to her about their sexual behavior (Freeman, 1989a, cited her as Mead's "very closest Samoan friend," p. 1018). Therefore, despite her contention during the interview of having lied in the past to Mead, we are supposed to believe that the resolution of this scientific controversy rests with her present "truthful" testimony. The film also claims that this was Mead's "chief informant," but according to Freeman (1989a), she was too old to have been one of Mead's actual informants (the informants Mead studied to evaluate her coming-of-age thesis were all between the ages of 8 and 20). She was apparently the same age as Mead at the time, so they simply developed a friendship (both she and Mead were born in 1901; Freeman, 1989a, p. 1018).

But, it is also unclear just how deep this friendship was. Mead wrote that Fa'apua'a principally resided in a village that was several miles from her research site. Mead (1928) acknowledges her as "the Taupo of Fitiuta" (p. viii; i.e., the ceremonial virgin of the village of Fitiuta on the opposite side of Ta'u from where Mead stayed) and seems to have spent time with her only during formal intervillage visits (*malaga*; Mead, 1977). Further, Fa'apua'a recently told a newspaper reporter that in 1925–1926 she did not know Mead was an anthropologist, although she also said that Mead "was always going around with a notebook and writing things on it" (Caton, 1990, p. 163; Freeman, 1989a). Mead's occupation is a rather important detail of which Mead's "very closest Samoan friend" would be unaware. Whether she really did not know at the time, or she had forgotten Mead's principal reason for being there, such disclosures further call into question this informant's credibility as a witness in the matter. In any event, the reliability of any testimony given some 60 years after the fact must be viewed with caution, especially if it supposedly settles a controversy in a scientific manner.

Aside from these problems, the fact is that Mead's book describes the sexual behavior of many of her informants and others in the villages she studied, not just Fa'apua'a Fa'amu's, so even if information about her is false, her current recanting does nothing to address the other material in Mead's book. Moreover, Mead's descriptions of sexual behavior are often based on information that most young villagers would have known at the time if they kept up with local affairs, so she would not have had to rely solely on personal accounts. Readers can evaluate this on their own, simply by perusing Mead's book.

If we focus on the scientific validity of this "crucial evidence," we find that no safeguards were used during this data collection to eliminate the possibility that the information obtained was contaminated by the method of collection.

Freeman (1991) constructed the events leading up to the interview to suggest that Heimans, the filmmaker, just stumbled upon her testimony. He wrote at the beginning of his article that "[w]hen he reached Fitiuta . . . Heimans was approached by a dignified Samoan lady, dressed as though she were going to church. She told him, through an interpreter, that she had a confession to make and would like to have it recorded on film so that all might know of it" (p. 104). Later in the article he claimed that "when she heard that an Australian filmmaker was coming to Manu'a, she decided, being a devout Christian, to make a formal confession" (p. 116).

This "serendipity" (Freeman, 1991, p. 104) stretches credulity, especially when it appears that Freeman was with the camera crew that recorded the "confession."[3] Thus, he was present, or at least nearby, before and during the interview, so it is possible that his presence, just 4 years

---

[3]This is deducible from references provided by Freeman. He gave the following reference for the interview discussed in his 1989a article: "Fa'apua'a Fa'amu 1987 Interviews with Galea'i Poumele and D. Freeman, on November 13. Fitiuta, Manu'a, American Samoa" (Freeman, 1989a, p. 1022). However, in a separate article, he provided a reference for the interview in the film: "Fa'apua'a Fa'amu 1987 Transcript of interview of 13 November with Frank Heimans of Cinetel, Sydney, Australia. Galea'i Poumele, Secretary of Samoan

after the publication of his controversial book, had some effect in terms of what she sensed was expected of her. In fact, according to Freeman (1991, p. 122), she had lived "in Hawai'i from 1962 to 1986," so it is also likely that she had heard about the controversy while living there, especially because it had been well-publicized by the U.S. media, including "the 'Phil Donahue Show' and the CBS morning news show" (Holmes, 1983a, p. 541). At the very least, a camera crew investigating the controversy must have heightened her expectations in some way.

But, more importantly, the consensus in the Ta'u area had been for some time that Mead was duped by mischievous informants, so local residents were well aware of a controversy over Mead's work, even without Freeman stirring things up. Being the last person alive from that era would mean that she would have been "fingered" by local residents as partly responsible for the embarrassing things Mead wrote about residents of the island. Thus, when the Heimans' film crew arrived, it is possible that she was pressured to make her "confession." Under these circumstances, not to deny that she had been "promiscuous" when young would have contradicted this consensus and perpetuated the embarrassment. Cranberg (1983a) described the consensus about Mead on the island where she conducted her study—note that this disdain for Mead predates the release of Freeman's book:

> Based on my one-year experience as a medical doctor on the island [during 1979–1980], it was surprising to learn that much of the wondering [about Mead] is petulant, irritated, and contentious. Many Ta'u natives today wonder why she should have earned generous royalties while reporting from freely given sources of information. . . . Some wonder why she did not wait to write her book until after all its characters had died. Far from being a hero on the island, she appears to be a source of embarrassment to its people. Mention of her name is a touchy subject, and her book is treated as if it were best forgotten quickly. (p. 182)

Although we could excuse such vagaries of data collection on the presumption that some circumstances are beyond anyone's control, when we examine the conditions under which the interview portrayed in the film was conducted, it is hard to overlook the strong possibility of other "interviewer effects." For example, the interviewer in the film was a High Chief and the Secretary of Samoa Affairs, Government of American Samoa. The fact that a High Chief who was also a high-ranking government official was interviewing her on a sensitive sexual subject in front of a camera

---

Affairs, Government of American Samoa, interviewer" (Freeman, 1989b, p. 761). Note that both interviews took place on the same day on the island of Ta'u. This is not evident in his 1991 article.

raises additional questions regarding how frank Fa'apua'a Fa'amu could be under those circumstances. Furthermore, it appears that the interviewer is the son of someone else who was also a young woman on Ta'u when Mead was there, and was friends with Fa'apua'a during that time. This is deducible from the transcript of the interview provided by Freeman (1989a, p. 1020), when Fa'apua'a Fa'amu said to Galea'i Poumele, "Then, your *mother* Fofoa and I would pinch one another and say: "We spend the nights with boys, yes, with boys!" (italics added). Thus, by implication she would have been charging that the mother of the interviewer had been (what is now defined as) "promiscuous" as well. And, finally, later in the film the interviewer expresses his strong negative opinions regarding Mead's book. Several scenes later in the film, the interviewer says: "I think [Mead] made up her mind that she's going to write that theory according to what she believed, but not according to what the people are living here."

Under these circumstances, why should this person admit on camera and in the presence of authority figures that she engaged in *what is now* considered immoral behavior? Quite likely, as word about Mead's book got back to her informants, who found themselves in an increasingly Christianized culture (cf. Gerber, 1975), they became upset and denied the "whole thing." Evidently, Fa'apua'a Fa'amu had been a devout Christian most of her adult life and her father had been a lay preacher (Freeman, 1989a, p. 1018). To admit to premarital sexual activity would constitute an admission of hypocrisy and immorality, as well as invite public disgrace in front of her community, children, and grandchildren (cf. Schoeffel & Meleisea, 1983). In fact, Fa'apua'a Fa'amu is on record as saying that "she was very displeased with Mead after she heard the things she said in her book" (Caton, 1990, p. 13).

Understandably, then, the young women described by Mead have likely never forgiven her for telling the world that they engaged in premarital sex, therefore shaming them as well as their families and villages. After all, there is no question regarding which villages she studied and there were only 25 postpubertal adolescent females in the villages at the time. Accordingly, in spite of her use of pseudonyms, many of the sexually active informants could have been identified by those familiar with the villages. With information from Mead's vivid case history presentations (e.g., family background, social reputation, and dominant personality characteristics), in conjunction with information in Appendix V of *Coming of Age* (which provides information such as age, family structure, and whether time had been spent in the "pastor's household"), the identity of many of her informants could surely have been recognized. In fact, with information from Freeman's (1989a) article, it was obvious to me who Fa'apua'a Fa'amu is in Mead's book. If I could identify her in this way, surely anyone from her community could have identified her and the others described by Mead. Incidentally, Mead's account of Fa'apua'a Fa'amu occupies less than one page of her book.

Thus, it is entirely plausible that it was far easier for Fa'apua'a Fa'amu to lie on camera and to say that what she said 60 years ago was a lie. It is also plausible that she now believes that what she told Mead was a lie. Sixty years of reflection can lead to all sorts of "mental acrobatics," ranging from repression to cognitive dissonance reduction (cf. Festinger, 1957). This would account for the fact that she has sworn on a bible to attest she is now telling the truth (Freeman, 1991, p. 116).

In view of these plausible alternative explanations for this testimony, there is ample reason to dismiss Freeman's (1989a) conclusion that "there is now an adequate, empirically based explanation of why it is that Margaret Mead's account of sexual mores and behavior in Samoa is radically at odds with the accounts of all other ethnographers" (p. 1021). On the contrary, his "evidence" can be seen in the context of a general desire of some Samoans to "deny" Mead's account of this period in their past (cf. Gerber, 1975; Schoeffel, 1986). There have been enough Samoans who have gone on record as being upset with Mead's "exposé" for this to be a plausible explanation for the discrepancy mentioned by Freeman, and for this "crucially important new evidence" to be called into question (cf. F. Wendt, 1984).

Finally, it is highly doubtful that Popper (1968), himself, would be impressed with the claim that this interview is key evidence in a refutation. In fact, from the following passage it is likely that Popper would dismiss it as a "non-reproducible single occurrence" and "a few stray basic statements:"

> We say that a theory is falsified only if we have accepted statements which contradict it. This condition is necessary, but not sufficient; for we have seen that non-reproducible single occurrences are of no significance to science. Thus, a few stray basic statements contradicting a theory hardly induce us to reject it as falsified. (p. 86)

In spite of all of these problems with this evidence, Freeman (1989b) declared that he has "claimed closure of the controversy" (p. 760) and that his refutation of Mead is "now fully vindicated" (Freeman, 1992, p. 2). But, this supposed "resolution" of the controversy represented by this interview with Fa'apua'a Fa'amu must be dismissed for lack of scientific rigor, and the film that claims to resolve the Mead–Freeman controversy appears to be little more than propaganda produced to support Freeman's "mission" to discredit Mead. It is ironic, therefore, that in the film Freeman states: "The scientific truth is something that cannot be settled politically. It's something that depends on the evidence" (Heimans, 1988).[4]

With this understanding of the political context of the controversy, we now turn to an examination of the specific evidence presented by Freeman in his claim that Mead's conclusion concerning her coming-of-age thesis was "preposterous."

FREEMAN'S EVIDENCE: ADOLESCENT STORM AND STRESS

The thrust of Freeman's statistical critique of Mead's coming-of-age thesis is based on three arguments. One is that she contradicted herself in her own work; a second is based on assertions that crime and delinquency are, and were, rampant in all parts of Samoa; and a third is based on his conversations with a handful of "highly educated Samoans."

Argument 1: Apparent Contradictions in Mead's Work

Freeman argued that Mead contradicted herself in her own work and that she really described a situation where adolescent storm and stress was more common than in Western countries. In pointing out this supposed contradiction, he noted that Mead identified several informants who were exhibiting delinquent behavior, and he posed the question "what can be concluded about delinquency among Samoan female adolescents from the information Mead herself has provided?" (1983, p. 255). Indeed, in a chapter entitled "The Girl in Conflict," Mead discussed in detail seven individuals who exhibited various forms and degrees of social deviance. Let us begin with an examination of Mead's definitions of deviance and delinquency, and how she applied these definitions to these seven informants.

Mead defined the *deviant* as someone who departs from the pattern of behavior prescribed by his or her group (p. 169). From this reference point, Mead noted two types of deviance in her sample: *downward* deviance and *upward* deviance.

---

[4]Those interested in reading the various commentaries about the Mead–Freeman controversy can consult a number of sources from both "sides" that: (a) support Freeman (e.g., Appell, 1984; Appell & Madan, 1988; Caton, 1984, 1990; Leach, 1983; Tribe, 1984); (b) criticize him (e.g., T. Baker, 1984; Brady, 1983, 1991; Ember, 1985; Goodale, 1984; Handler, 1984; Kuklick, 1984; Laing, 1987; Leacock, 1988; Murray, 1990, 1991; Nardi, 1984; Patience & Smith, 1986; Paxman, 1988; Rappaport, 1986; Reyman, 1985; Scheper-Hughes, 1984; Schoeffel & Meleisea, 1983; Shore, 1983; Turnbull, 1983); and (c) both support and criticize him (e.g., Badock, 1983; Buchholz, 1984; Hooper, 1984; Muensterberger, 1985).

The downward deviants in Mead's (1928) view were *delinquents*. She used this term "to describe the individual who is maladjusted to the demands of her civilization, and who comes definitely into conflict with her group, not because she adheres to a different standard, but because she violates the group standards which are also her own" (pp. 171–172).

The upward deviants, on the other hand, were stimulated to move "beyond" their culture resulting in the "wish to exercise more choice than is traditionally permissible" (p. 171). This desire to make nontraditional choices was "encouraged by the educational system inaugurated by the missionaries" (p. 171), so it would therefore have been less common in traditional Samoan culture. Of the seven deviants in Mead's sample, three were classed by her as upward deviants (p. 171; their pseudonyms are Ana, Lita, and Sona). It is important to note that Mead concluded that these three individuals were managing adequately within their community by utilizing the options available to them.

The other four deviants were not so easily classifiable, as the following passage reveals: "were there really delinquent girls . . . girls who were incapable of developing new standards and incapable of adjusting themselves to the old ones? My group included two girls who might be so described . . ." (Mead, 1928, pp. 172–173). Thus, only two informants clearly fit Mead's definition of delinquency. Of these two, one (Mala) had just reached puberty, whereas the other (Lola) was 2 years past puberty. But Mala had exhibited behavioral difficulties since early childhood, while Lola had done so for at least 4 years (pp. 177–179).

The other two deviants were marginally delinquent: One (Sala) had not yet encountered "serious conflict with her community" except that she had been expelled from the pastor's house for "sex offenses" (p. 181); and the other (Siva) was a "delinquent in the making" who was only 11 years old and had not yet reached puberty.

The main problem with the inferences Freeman drew from these qualitative descriptions is that he made inappropriate and inexact statistical comparisons with other cultures. These comparisons are based primarily on statistics from Great Britain, which Freeman argued indicate that the "delinquency rate" in Ta'u would have actually been "about ten times higher than that which existed among female adolescents in England and Wales in 1965" (p. 258). This startling assertion is based on the claim that if "among the twenty-five adolescents [Mead] studied there was *one* delinquent act per annum, this is equivalent to a rate of forty such acts per thousand" (Freeman, 1983, p. 257). Note that all that he did here was to use the algebraic equation "$1/25 = x/1,000$," where "x" equals 40 upon solution ($1,000/25$), and to cite the statistic that the delinquency rate in England and Wales in 1965 was 4.00 per 1,000.

Freeman claimed that this statistical sleight of hand is a "conservative" assumption, but it is actually a specious one because not only is he inappropriately extrapolating from a small sample to a large one, but he is setting up a "zero-tolerance" standard of deviance that no culture can meet. In other words, by his reckoning, only 1 of Mead's 25 informants had to commit a delinquent act to conclude that Samoan culture is rife with delinquency, because the resulting statistical rate would indicate ten times the delinquency rate of the Western countries compared.

But, add to this specious argument the fact that Freeman's statistics are for *indictable offenses* in England and Wales, and there is little left to his claim that Mead contradicted herself. This use of these indictment statistics ignores the well-established finding that far more crime takes place in Western societies than is reflected in such statistics. For example, in the criminology literature it is known that:

> not all offenses result in a charge being laid; and not all charges result in a conviction. Indeed, only about one known victimization in 100 comes to a conviction, if attention is confined to crimes against persons such as assault or rape, only one victimization in ten is reported to the police and only one such report in ten results in a conviction. (Tepperman, 1977, pp. 7–8)

Thus, Freeman's statistical argument suggests something close to the *minimum* estimate for one culture (Western), whereas Mead's account would represent something close to the *maximum* estimate for another (Samoan). Moreover, her cases involve mainly the violation of social mores that often do not lead to indictments in Western countries (cf. Young & Juan, 1985). The fact that a minority of her sample engaged in some delinquency stands in stark contrast to Gold and Petronio's (1980) observation that more than "80% of American adolescents admit to committing one or more delinquent acts ... in the course of a few years of adolescence" (p. 523).

Using Gold and Petronio's figure of 80% to establish a more appropriate estimate for Western countries (their figure summarizes research from the 1960s and 1970s, whereas Freeman's estimate is taken from the 1960s), we can evaluate various estimates from Mead's sample. For example, we could count the two informants who had minor delinquency problems (Sala and Siva) along with the two Mead more clearly classed as delinquent (Mala and Lola). Even so, including all of Mead's 50 informants would mean that only 8% of her sample engaged in some form of delinquency. Note as well that this estimate includes all causes, none of which appear to have had their genesis after puberty. We can arrive at another, more liberal, estimate if we count only the 25 informants between 15 and 19 years of age. (This is the age range Freeman uses for delinquency statistics from Great Britain.) This produces a rate of 12% (3/25), because only three of Mead's informants were "delinquent" or "potential delinquent" in this age range (recall that Siva was only 11). Clearly, these figures are much lower than the 80% estimate that applies to U.S. society.

Freeman (1983) introduced another, equally problematic argument, when he implied that Mead's "four delinquents and three 'upwards deviants,' . . . together, make up 28 percent of her sample of twenty-five female adolescents" (p. 258) and should be considered to have experienced adolescent storm and stress. He did not specify any means of confirming these as cases of storm and stress. Instead, he simply claimed that Mead engaged in a "decidedly unscientific maneuver" by not pursuing this issue. Freeman did not pursue the issue of how and why these cases should be thus classified; instead, this argument is tagged onto the argument regarding crude delinquency statistics.

Had he wanted to ensure that he was being "decidedly scientific," he should have sought ways to operationalize storm and stress, and he could have done this by examining the contemporary literature that directly assesses the difficulties that Western adolescents can encounter in coming of age. He would have found that this literature reveals rather high levels of such difficulties. For instance, in their longitudinal study of a sample of U.S. male adolescents, Offer and Offer (1975) classified only 23% as experiencing a conflict-free, "continuous" growth, that would be comparable to that described by Mead. Although only 21% of their sample were classified as experiencing what they termed *tumultuous* growth—severe turmoil—another 35% were classified as experiencing *surgent* growth—moderate difficulties in emotional adjustments. Thus, some 56% of their sample experienced a variety of conflicts and emotional difficulties, ranging from moderate to severe. Similarly, other researchers have found that a majority of U.S. young people at some time have problems in terms of developing and sustaining a sense of commitment and belief (e.g., Marcia, 1980). In addition, up to two thirds of those in samples of Canadian young people have experienced problems with parental conflict, a lack of a sense of direction in life, and an uncertainty of belief (Côté, 1986; Pfeiffer & Côté, 1991).

If we return to Mead's account of her cases of "girls in conflict," we can gain a better sense of what these cases mean for her coming-of-age thesis. First, the two cases that Mead actually classified as delinquent were not exhibiting symptoms of an adolescent storm and stress somehow associated with puberty, because one informant had exhibited behavioral difficulties since early childhood, and the other had done so for at least 4 years but was only 2 years past puberty. Thus, they exhibited what were likely problems of childhood that persisted into adolescence (Freeman appeared to consider adolescence the postpubertal period). In fact, Mead (1928) did not report any serious behavioral problems that began after puberty, with puberty defined as the onset of menstruation (pp. 283–284).

It is also important to note that, despite Freeman's charge that Mead was a cultural determinist, Mead actually implied that an important cause of the problems faced by her two delinquents was biological, or in her words "temperamental." As we see from the following passage, Mead argued that their problems stemmed from an incompatibility between their temperaments and the constraints of Samoan culture:

> Lola and Mala both seemed to be the victims of lack of affection. They both had unusual capacity for devotion and were abnormally liable to become jealous. Both responded with pathetic swiftness to any manifestation of affection. At one end of the scale in their need for affection, they were unfortunately placed at the other end in their chance of receiving it. Lola had a double handicap in her unfortunate *temperament*. . . . Her *temperamental defects* were further aggravated by the absence of any strong authority in her immediate household. . . . So it would appear that their delinquency was produced by the combination of two sets of causal factors, unusual emotional needs and unusual home conditions. Less affectionate children in the same environments, or the same children in more favourable surroundings probably would never have become as definitely outcast as these. (p. 180, italics added)

Ironically, this is the type of "interactionist" thinking that Freeman claimed is lacking in Mead's analysis. In reality, Mead made note of the limitations of the range of temperaments Samoan culture could accommodate (see especially Mead, 1930/1969, pp. 80–86; see also Mead, 1928, pp. 127-130, 247). Although Mead did not provide a definition for her usage of the term *temperament* in *Coming of Age*, she did so in a later publication (Mead, 1937), stating that she used

> the word temperament in the accepted technical sense for those aspects of the personality which are physiologically "given," as opposed to *character*, the latter being that part of the individual personality which is the result of the interaction between native equipment—or temperament—and cultural conditioning. (p. 558)

In other words, Mead was operating under the assumption that individuals have certain, sometimes idiosyncratic, genetic dispositions which they take with them into social situations. This is the meaning that is implied in *Coming of Age*, yet Freeman still felt justified in calling her a cultural determinist.

Contrary to the mythology that Mead painted a simple, idyllic picture of life in Samoa, the fact is that throughout *Coming of Age* Mead described various social and personal unpleasantries she witnessed while there. Included in these descriptions were instances of adjustment problems experienced by 7 of her 50 informants. Rather than constituting examples of adolescent storm and stress, as Freeman claimed, they more closely resemble conflict-produced adjustment problems that might be found in any culture. The fact that Mead documented these cases so carefully attests to the credibility and thoroughness of her research. Had she really been intent on presenting a distorted image of Samoan adolescence, she surely would not have devoted a full chapter to these problems. Instead, she showed us how individuals with certain temperaments and experiences found it difficult to come to terms with the constraints imposed by Samoan culture.

In evaluating this aspect of Freeman's "refutation" of Mead's work, therefore, it is clear that his evidence is itself easily refuted and his argument is poorly conceived: He did not provide definitions for the basic but controversial concepts of adolescence or storm and stress; he did not offer a theory that links social deviance or delinquency with adolescent storm and stress; and he did not evaluate Mead's research against the vast repertoire of theory and research on youth and adolescence that has accumulated since she carried out her pioneering work.

Argument 2: Delinquency Rates

Freeman presented another statistics-based argument in his attempt to refute Mead's claims that the adolescence she witnessed was one of relative ease. He first cited data that he said he collected in Western Samoa, documenting 746 convictions for crimes of violence committed by persons age 12 to 22 between 1963 and 1965 (he did not explain the methodology he used in collecting these data). In discussing these data, he stressed that the peak age for the convictions was 16, but he did little more than to imply that there is some sort of biological cause for the association between age and conviction rate. This implication is delivered with a citation of Katchadourian's (1977) *The Biology of Adolescence* as evidence that "the attainment of puberty is marked by steady and rapid improvement in physical strength, skill and endurance, and this development is also marked by involvement of adolescents in aggressive encounters of various kinds" (Freeman, 1983, p. 260).

Freeman continued with a presentation of more statistics that he said he compiled from Western Samoan police records. The offenses include: "assault and other crimes of violence; the 'provoking of a breach of peace'; theft and other offenses against property; trespass; rape and indecent assault; abduction; obstructing the police; uttering threatening insulting or indecent words; drunkenness; and perjury" (pp. 264–266).

He then compared these figures with those published by Sir Cyril Burt in 1925 for crimes in England. Burt's figures exhibit similar characteristics, especially a peak rate at age 16 with the highest rates occurring during the ages 15 to 19 years. From these comparisons he concluded that "it is clearly evident that the adolescent period in Samoa, [is] far from being 'untroubled' and 'unstressed' " (Freeman, 1983, p. 268).

When carefully scrutinized, however, this impressive statistical evidence argument loses its force rather dramatically, in terms of (a) a misrepresentation of Samoan culture and history (especially as it applies to the island of Ta'u), (b) methodological vagueness, and (c) vague application of human developmental theory.

Samoan Culture and History. From the point of view of Samoan culture and history, most of the crimes Freeman listed are irrelevant to communities based on communal sharing where there was virtually no private property or consumer items, no local police or courts, and no alcohol, as was the case before contact with the West. When applied specifically to the island of Ta'u in the 1920s, Freeman's case is weak. In fact, much of his overall critique rests on the tenuous assumption that Ta'u in 1925 would not have been significantly different in terms of day-to-day life from Western Samoa in the 1960s. He contended that all of the islands in the Samoan archipelago have been plagued with high rates of crime and delinquency for some time.

This contention is problematic given the historical, geographical, and political differences among the Samoan islands (discussed in more detail in the next chapter). For example, Western Samoa comprises two large islands, which in 1965 had a population of about 131,000 living in over 300 villages and one small city, compared with Ta'u in 1925–1926 comprising four villages with a population of about 1,200. In effect, then, Freeman claimed that the small island of Ta'u, with its modest population living in closely knit villages, was as crime-ridden as Western Samoa in 1965 and England in 1925. Holmes (1983b), who is discussed more in the next chapter, took Freeman to task on this issue early in the controversy:

> Freeman should have compared Ta'u with Ta'u. When I visited there in 1954, I found little of the deviant behavior he describes, and I can only assume that it was even rarer in 1925. William Green, the principal of the government school in American Samoa, in 1924, observed, "There has been no murder case in American Samoa since our flag was raised in 1900. Natives will suffer indignities for a long time before resorting to a fight." (p. 17)

Figures 2.1 and 2.2 give readers an idea of the relative size of islands in the Samoan archipelago.

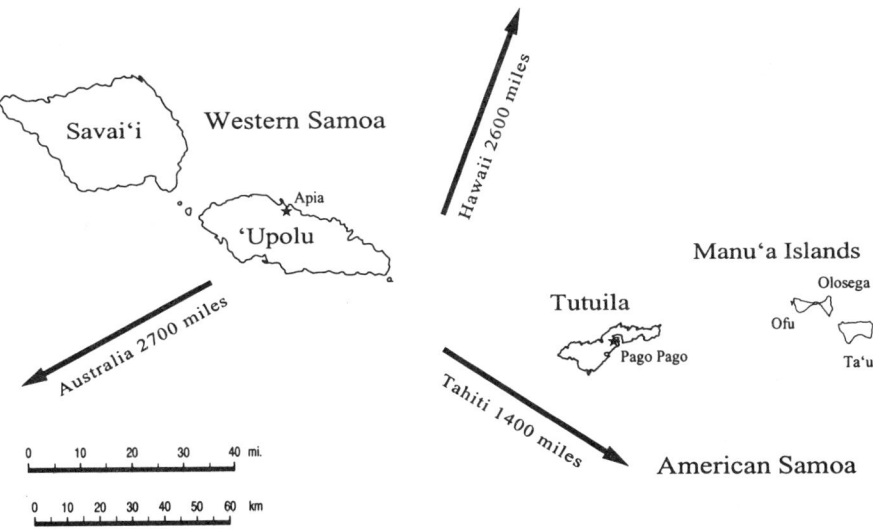

FIG. 2.1. The Samoan Archipelago.

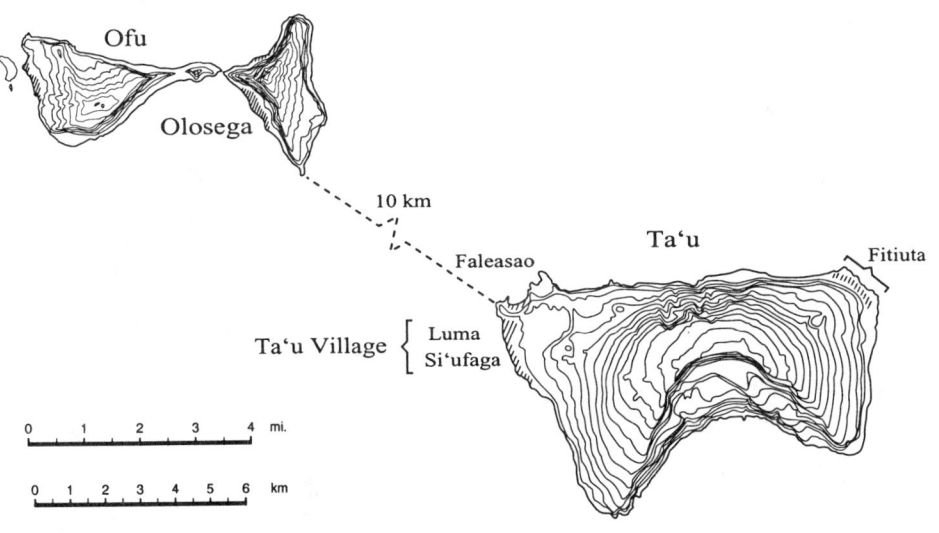

FIG. 2.2. Manu'a Islands

It is difficult to fathom how Freeman could ignore the element of cultural change with this argument, even in Western Samoa. For instance, O'Meara (1990) wrote that a major change over just one generation in the Western Samoan village he studied was the incidence of theft. In contrast to Freeman's claims, he noted that a

> generation ago or more theft was uncommon or even rare. . . . Today nothing is left in the plantations unless it is well hidden, and the crops themselves are frequently pilfered. When the family goes to church or to some other village affair, one member must always stay behind to . . . "guard the house" (pp. 69-70)

Interestingly, Freeman endorsed O'Meara's book as being "empirically exact" (cited by Spindler & Spindler, 1990).

Holmes is especially critical of Freeman on this point, noting that "nearly 70 per cent of the reported crime is said to be committed among the 18 per cent of the population which lives" in Apia—the only city in Western Samoa (Holmes, 1983c, p. 4). In another article Holmes (1983b) expanded on this:

> While Freeman is correct that Samoans share a common culture throughout the archipelago . . . [t]here is a great deal more criminal and deviant behavior in urban centers, such as Apia or Pago Pago, than in the smaller, more isolated communities in remote parts of Savai'i or the Manu'a Islands. In metropolitan areas, family heads and village councils of chiefs no longer hold control, and young men who migrate from outer villages to these areas, seeking work and excitement, often become intoxicated with their new-found freedom and behave in very non-traditional ways. (p. 17)

Methodological Vagueness. Freeman (1983) provided no discussion of his methodology other than to say that a "method was open to [him] . . . to compile from police records, a *random* sample of convicted offenders . . . The sample [he] compiled in this way totaled 2,717 convicted offenders" (p. 264), The vagueness of Freeman's account regarding his methodologies makes it difficult for other social scientists to evaluate his claims, especially given his contention to be "scientifically" refuting Mead. For example, he said these figures were for offenses committed in "the early 1960s, predominantly by inhabitants of the island of Upolu" (p. 266; elsewhere he said "c. 1963-1965," p. 263). The population of Western Samoa 15 years of age and over was about 63,937 in 1966 (Department of Statistics, Western Samoa, 1989, p. 3). This suggests that at least 4.25% of the adult population had been convicted of a criminal charge during the early 1960s (i.e., 2,717 divided by 63,937). Moreover, because he spoke of a random sample, we may assume that the actual population of convicted offenders from which he drew his sample must have been many times larger than 4% of the population, otherwise why bother to draw a sample, let alone a "random" one? How much larger is this population of convicted offenders? Are we talking about a society of ex-cons? I think not.

This vagueness is a problem throughout the book. For example, at one point he stated that "in the course of [his] fieldwork [he] observed fifty-six individuals aged 19 and under being physically punished by a parent, older sibling, or other senior member of a family . . . seventeen [of whom] were between the ages of 11 and 19" (p. 260). In presenting this as social scientific evidence, he should have supported it with, for example, accounts of (a) the period of time comprising his observations, (b) the contexts in which the incidents occurred, (c) the number of similar incidents that did not evoke punishments, and (d) comparisons with other cultures.

Human Developmental Theory. Finally, from the perspective of theories of human development, to say that physical development is "marked by" aggressive encounters is both vague and problematic, as is the suggestion that the "attainment of puberty" is related to these aggressive encounters. Had Freeman consulted the available literature on the topic before publishing his book, he would have encountered a literature review which concludes that "[a]n explanation of . . . delinquent behavior in terms of the onset of puberty seems weak . . ." (Gold & Petronio, 1980, p. 523).

In all likelihood, the conviction statistics he cited are related to the social structure of adolescence in 1960s Western Samoa. As we see later in this book, he should be aware that in precontact Samoan society, males would have undergone their rites of passage (tattooing and entry into the *'aumaga*—the group of untitled men) at age 16 and would be subsequently engaged in a meaningful participation in Samoan society (e.g., Turner, 1861/1986). By the 1960s, however, as economic, religious, and educational forces continued to transform Samoan society, many males no longer had the opportunity to experience these rites of passage. With no substitute for them, the traditional basis for developing a sense of identity was undermined and the transition to adulthood was becoming increasingly ambiguous and prolonged.

Sociologically speaking, these changes have resulted in a "cultural disenfranchisement" of Samoan youth. Having dislocated the young in the continuity of generational inheritance, it has become increasingly necessary for Samoan society to impose Western-style social control mechanisms to suppress and contain the disaffection among the young. The hostility and resentment among the young in response to the imposition of arbitrary adult authority is captured by Freeman in an anecdote he presented to support his case against Mead:

> From early adolescence onward Samoan youths may be observed grimacing and making threatening gestures at their elders, including chiefs, behind their backs, especially after having been punished or reprimanded; *with the attainment of puberty*, youths will occasionally lose control and openly attack those in authority over them. For example, in April 1965 a 31-year-old chief, patrolling a village of Savai'i to enforce the ten P.M. curfew, came upon a group of five male adolescents *who were breaking this curfew by playing a guitar and singing*, and he at once set about chastising them with a board. Instead of scattering, as would children, at this show of chiefly authority, one of these youths hurled a stone at the chief with such a force as to expose the bone of his forehead and put him in hospital for a fortnight with a concussion. (Freeman, 1983, pp. 261–262, italics added)

Note how Freeman associated "the attainment of puberty" with this incident of aggression. But numerous questions are left unanswered by this anecdotal evidence: We do not know how reliable Freeman's account is; we do not know how common such incidents are; and we do not know the actual age of the individual who allegedly threw the stone. Furthermore, is the behavior of the stone-thrower really attributable to the "attainment of puberty?" If so, how? What are the physiological mechanisms involved? Do they constitute proximal or distal causes? As part of his chapter on adolescence, Freeman cited crime statistics for persons ranging in age from 12 to 22. Are we to believe that all such individuals are "afflicted" with an inability to control their aggressive impulses against authority? If all are not afflicted, what saves those who are not? Freeman did not address any of these concerns.

If we take Freeman's word that his statistics and anecdotes are reliable, then at least part of their explanation must include factors associated with the social disorganization affecting the time and place of their origin. When we do so, we can consider the possibility that social factors constitute "proximate" causes of deviant behavior, and biological factors constitute potential "distal" causes of deviance (cf. Harris, 1983). Accordingly, without an immediate provocation, various biologically based potentials such as anger or aggression are not triggered, even if a particular individual is more prone to such reactions than is another individual. Such biologically based potentials do not operate at random or in a void, except in the cases of serious medical or organic psychiatric disorders, and adolescence is not a disorder.

To provide an example, the imposition of curfews is a postcontact social development—a social control mechanism that became necessary in certain areas to keep the now culturally disenfranchised and frustrated young person under supervision. In this light, we can understand Freeman's reports of the resentment and aggression on the part of some young people to be triggered by the imposition of arbitrary adult authority. Moreover, Western influence introduced the notion that childhood and adolescence can be periods of idleness and leisure. This notion of youth as a time of leisure stands in contrast to the well-documented fact that young people in precontact Samoa (and most agrarian societies) played an integral role in the division of labor. Therefore, it is not surprising that in the modern era some Samoan youth would now experience their life as one of "servitude" (Freeman, 1983, p. 259). In chapter 4, I examine the precontact circumstances that would have made adolescence a more rewarding period of life than it now is.

Given Freeman's own call for an "interactionist anthropology" that considers the interaction between culture and biology (e.g., Freeman, 1987c), he must be willing to entertain the question of how the social environment can produce biological reactions such as aggression. The deleterious effects of this social disorganization on Samoan youth have been analyzed by several social scientists (P. Baker, Hanna, & T. Baker, 1986; Leacock, 1987; MacPherson & MacPherson, 1985; O'Meara, 1990; Pitt & MacPherson, 1974; Yusuf & Peters, 1985). They have also been characterized in A. Wendt's (1973, 1977, 1979) fictional accounts of Western impact on Samoan society and Samoan youth.

Argument 3: Expert Samoan Opinion

Freeman's third piece of evidence regarding adolescence is clearly the weakest. This evidence comes from his account of casual conversations he had with several Samoans. Normally, this type of evidence is inadmissible as scientific evidence. Undoubtedly, even a court of law would dismiss it as "hearsay." Nevertheless, it must be examined here because some observers have apparently been swayed by it (e.g., Muuss, 1988, p. 142, wrote that this piece of evidence "devastates Mead's conclusions").

Freeman (1983) stated that he had never met "a Samoan who agrees with Mead's assertion that adolescence in Samoan society is smooth, untroubled, and unstressed" (p. 259). To substantiate his claims, he cited conversations with four "highly educated Samoans" and an unspecified number of Samoan adolescents whom he and his wife "came to know particularly well" (p. 259). From these informal sources he felt justified in concluding that "it is clearly evident that not a few Samoans, during adolescence, are subjected to psychological stress'' (p. 260).

Although it is impossible to verify Freeman's "data" from most of his "informants," one "informant" wrote a review of his book, and certainly did not endorse it. In the context of Freeman's claim, her comments are quite illuminating:

> I have seen days (and nights) like Margaret Mead's and moments of mayhem like Freeman's. No one who has lived in Samoa long could doubt the existence of both. My only problem is with people who, like the blind men and the elephant, feel for one aspect or another and draw conclusions about what Samoans really "are." (Ala'ailima, 1984, p. 91)

Freeman also cited this source as having recorded "that Samoan adolescence is . . . 'a period of 'Sturm und Drang' " (Freeman, 1983, p. 259). However, in her lighthearted book describing some adventures she had adjusting to Samoan life, she mused that Mead must have been wrong because of several experiences she had with young people in her charge (Ala'ailima, née Calkins, 1962—she was an American who married a Samoan in the 1950s). It is ironic for Freeman's argument, however, that most of the incidents she described involve the attempts of these young people to engage in premarital sexual activity without their adult custodians finding out.

More to the point, however, even if a number of Samoans are of the opinion claimed by Freeman, so what? No one in Samoa is an expert on adolescence in 1920s Ta'u as far as I know and Freeman named none. To put this in another context, would it carry any weight if I went to a particular area and asked four people what they thought about something, and then reported their opinions to the social science community? I think not.

Although I am reticent to engage in hearsay, to make my point regarding the reliability of opinion, I informally asked similar questions in Samoa, but got different answers than did Freeman. The consensus of these answers was that young people in contemporary Samoa are no different from young people elsewhere. There is a sense of ongoing change, however, but this should come as no surprise given the extent of Western influence, especially in American Samoa, which has strong ties with Hawaii and California. In American Samoa, there seems to be a growing concern about the importation of gang mentalities from these U.S. centers, and an increase in violence as a result. In Western Samoa, the concern seems to growing about the lack of jobs available to the young when they graduate from high school and the inability of some young people to find a place in Samoan society because of recent economic and social change. These problems are explored in chapter 7.

LIES, DAMNED LIES, AND STATISTICS[5]

As we saw previously, Freeman's case against Mead's portrayal of Samoan culture is partly based on comparisons of Samoan crime statistics with those from other countries. In presenting his case against Mead, he also presented statistics pertaining to all ages, not just adolescents. It is with these statistics that he made his most forceful arguments about the "darker side" of Samoan culture. In doing so, however, he has left the impression with some readers that Samoans are "a race of sex-starved rapists" (Levy, 1984), or that they are a "violent . . . delinquent, rape and suicide prone . . . people" (F. Wendt, 1984, p. 95). Marshalling this statistical evidence, Freeman seems to have swayed a number of people to his side, creating the impression that Mead was a fool for not having noticed these "obvious" traits and behaviors among Samoans. But, let us step back for a moment and consider his statistics in a different context.

Most of the statistics used by Freeman are crude rates, in which crimes are counted per year against a population denominator of 100,000. For instance, from his own research he stated that

> In 1966, when the total population of Western Samoa was about 131,000, the number of forcible and attempted rapes reported to the police in Western Samoa was thirty-eight. This is equal to a rate of about sixty rapes per 100,000 *females* per annum, a rate twice that of the United States and twenty times that of England. (pp. 248-249, italics added—note the rate is per female)

Setting aside the potential problems with the accuracy of Freeman's statistics for a moment, we must ask what these statistics say about what life was/is like for the average Samoan female, particularly one coming of age? Samoan villages range in size from 200 to 500 (Meleisea, 1987b, pp. 5-6), so we will assume that the average village has a population of about 350.[6] Using basic algebra, when we convert the overall probability of $60/100,000$ to a specific probability of $x/175$ (assuming that roughly one half the population of a village was/is female), we find that there was a probability of .105 rapes or attempted rapes being reported in a given village in Western Samoa during 1966. In other words, there may have been one reported incident of rape or attempted rape for every 10 villages, over an entire year! The intention of stating these figures in this

---

[5]This phrase is taken from Mark Twain (1924, p. 246) in reference to his being perplexed by the use of statistics.

[6]A feature of Samoan social organization is that a village will not grow beyond a certain size without out-migration taking place. Presumably villages do not grow much larger than the 500 figure cited by Meleisea, although O'Meara (1990) cited 600 as the upper limit.

way is not to diminish the severity of rape as an offence, but one is left with a much different impression of daily life in Samoan culture.

Freeman nevertheless felt justified in alleging that "rape is unusually common in Samoa" (p. 249). Although he did say that he is aware of the problem of engaging in the cross-cultural comparisons of such statistics (p. 164), I am not convinced that he has taken these problems seriously. Clearly, he placed no *scientific* restraints on his comparisons. For example, it is very difficult to compare cultures in terms of the rate at which a crime like rape is reported to police. In Canada, for instance, it has been estimated that only 1 in 10 incidents of rape is reported to the police (Tepperman, 1977). Freeman contended that "in Western Samoa a very considerable proportion of forcible and surreptitious rapes are, in fact, not reported to police." He does not, however, give us any evidence for this assertion or how this rate of reporting compares with other countries.

Returning to Mead's study, based on Freeman's rates, using the population figure of 600 for the three villages she studied (Mead, 1928, p. 260) and dividing it by 2 to estimate the number of females, there would have been a probability of .18 rapes or attempted rapes being reported in the three villages she studied on Ta'u during 1925–1926. In other words, using Freeman's figures, one would expect a report perhaps once every 5 years. Because Mead was there for only about 6 months, the probability of her encountering an incidence of rape or attempted rape would be one half the figure just cited. Accordingly, using Freeman's statistics, there was a statistical probability of only .09—a 1 in 10 chance—of a rape or attempted rape being reported while she was there.

But, add to these considerations differences in time and place, and the probability of an occurrence diminishes even further. For example, Freeman vaguely referred to missionary accounts of rape, and accounts of rapes during the first three decades of the 20th century (p. 249). However, Rowe (1930) provided the following account of a rape in Apia, Western Samoa, during the 1920s:

> A circumstance which created a considerable sensation about the end of 1922, and ruffled a period of calm, was that one of the white sisters from the hospital was raped by a Samoan on the lonely stretch of up-hill road. . . . *It was an occurrence, I think, entirely unprecedented in Samoa.* (p. 133, italics added)

Rowe's (1928) account is totally consistent with Mead's own statement that "[e]ver since the first contact with white civilization, rape, in the form of violent assault has occurred occasionally in Samoa" (p. 93). Freeman (1983) cited Mead as writing that "the idea of forceful rape . . . is completely foreign to the Samoan mind" (p. 244), but this is taken from Mead's (1931) article "The Role of the Individual in Samoan Culture" in which she explicitly restricts her generalizations about Samoan practices "to refer to Manu'a specifically" (p. 545).

Holmes (1983c) said the following about Freeman's rape statistics:

> The statistics which Freeman maintains could not be obtained until 1981 [after Mead's death] reveal the fact that in American Samoa some fourteen rapes per year occurred from 1975 to 1980. Freeman also states that court records from American Samoa show 'numerous cases' of rape recorded for the first three decades of this century. . . . I know for a fact that there are no records of rapes occurring in Manu'a at the time of Mead's research and none for 1954 when I was doing my restudy of Mead. (p. 4)

Instead, Holmes attributed occurrences of rape largely to the Naval personnel stationed at Pago Pago and to other Western/urban influences—to "social disorganization brought on by cultural change and not a biological disposition toward aggression inherent in the Samoan make-up" (p. 4).

Finally, although Freeman (1983) cited Rowe elsewhere, he conveniently omitted Rowe's reference to how unusual rape was. In view of these arguments, although rape undoubtedly occurred and occurs in Samoa, as in most other countries in world, Freeman appears to have blown its significance out of proportion in his quest to discredit Mead. Unfortunately, he also simultaneously discredited Samoan culture.

This technique of evaluating Freeman's statistical arguments can be applied to the other statistics he presented regarding murder, assault, and assault causing bodily harm.

For murder, his estimate of 6.6 per 100,000 in Western Samoa in 1977 translates to a probability of .0231 murders in a given village of 350 in a given year. Therefore, there might have been one murder for every 50 villages that year. My own check of Western Samoan records reveals two murders reported in 1964 and one in 1965 (Police and Prisons Department, Western Samoa, 1965). Although Freeman used data from the mid-1960s for his other comparisons, he curiously did not do this for this crime, despite the fact that the report just mentioned is available at the Nelson Library in Apia, Western Samoa. The fact is that this crime shows great fluctuations there. For instance, Freeman reported 10 murders for 1977, but more recent statistics reveal only 6 murder charges in 1985, 7 in 1986, 2 in 1987, and 7 in 1988 (Department of Statistics, Western Samoa, 1989, p. 81).

For "affrays," or feuds that erupt into violence between families, Freeman cited a statistic of 40 per 100,000, which translates to a probability

.14 affrays per village of 350 in a given year. Therefore, there might have been one affray reported for every seven villages that year.

Finally, Freeman's figures from the mid-1960s in Western Samoa for "assault causing bodily harm" translate from 105.1 per 100,000 to .37 per village of 350 per year. Those of common assault translate from 773 per 100,000 to 2.71 per village per annum. Thus, there might have one serious assault reported for every three villages that year, and about three minor assaults per village.

Consistently, when we analyze Freeman's statistics by applying their probabilities, we get a much different picture of daily life than the bleak one painted by him. Clearly, using his own statistics, the average Samoan neither engaged in violence nor would have been in daily fear of it. Yet, Freeman has succeeded in swaying large numbers of people to his dark view of Samoan culture. His eloquent prose together with his use of anecdotes of incidents that took place over great amounts of time makes for dynamic reading, but not for good science.

Undoubtedly, tempers flare and violence occurs in Samoa, but to hold Samoa up as an extreme example is highly misleading to the world, and highly insulting to Samoans. In his attempt to discredit Mead for not emphasizing these things in her research, Freeman also disparaged Samoan culture by making it sound like Samoans beat on each other as a daily practice.

Finally, in reference to Mead's coming-of-age thesis, when we review Freeman's statistics in the above light, there is nothing in them that places Mead's conclusion in question.

CONCLUSION

To conclude this investigation of Freeman's case against Mead, in the final analysis, Freeman's "refutation" is itself easily refuted. His criticism that Mead ignored biological factors is not only erroneous, it is empty, because he himself did not suggest ways in which she might have examined such factors any more than she actually did. His evidence concerning storm and stress among Samoan adolescents is weak and can easily be interpreted in ways other than the one he has chosen as part of his quest to discredit Mead. Most problematic is his contention that Ta'u of the 1920s can be understood with observations and statistics collected elsewhere during the 1960s. Further, Freeman did not examine the issues involved from the appropriate social scientific perspectives. Had he done so, and examined human development and sociological literature, he would have developed a more sophisticated

CONCLUSION

position on the matter of adolescent storm and stress. Therefore, he could have considered the plausibility of Mead's formulations in light of the dramatic social and economic forces that have affected Samoan culture during the 20th century. Finally, Freeman's use of crime statistics seriously distorts Samoan culture, and has few implications for Mead's coming-of-age thesis.

CHAPTER THREE

Mead's Culpability

Now that we have reviewed the heated controversy stimulated by Freeman's book and evaluated Freeman's case against Mead, we can move toward a "verdict" on Mead's coming-of-age thesis. In this chapter, we look beyond the arguments presented by Freeman and examine Mead's coming-of-age thesis based on several independent sources of evidence. This "cross-examination" begins with the "testimony" of expert witnesses, none of whom have any apparent political axe to grind. Following their testimony, we inspect the only available "material evidence" in the case. Finally, we review what Mead said in her own defense, in response to early criticisms of her book.

EXPERT WITNESSES

Lowell Holmes

The first witness is Lowell Holmes. Holmes is an American anthropologist whose doctoral thesis was a restudy of Ta'u during the 1950s. For almost 40 years, his career has been devoted largely to the study of Samoan culture, particularly in the Manu'an group of islands (see especially Holmes, 1957a, 1957b, 1974, 1980a, 1987; Holmes & Holmes, 1992). No other anthropologist has studied Manu'a as extensively. Thus, Holmes is the most qualified anthropologist to consult in this case.

Based on his research, Holmes has no hesitation endorsing most of Mead's findings. When the controversy first erupted, Holmes (1983b) spoke in her defense, saying that despite the great possibilities for error in a seminal scientific study, [he] found Mead's Samoan research remarkably reliable. The differences between [their] findings that could not be attributable to change were relatively minor and, in most cases, involved *not discrepancies in data but differences in interpretation* . . . (p. 16, italics added)

In *Quest for the Real Samoa*, Holmes (1987) confirmed that he found "the validity of her Samoan research remarkably high." Moreover, he can "confirm Mead's conclusion that it was undoubtedly easier to come of age in Samoa than in the United States in 1925" (p. 103; see also Holmes & Holmes, 1992, pp. 143-144).

In evaluating Holmes' credibility here, it is most interesting that during the 1960s he advanced several criticisms of Mead's work, and was consequently embroiled in a public controversy with her (e.g., Mead, 1961, 1930/1969). As he explained, he therefore has no reason to rush to her defense, but he also saw no reason to deny the validity of the bulk of her work, or to personally discredit her. In fact, Holmes (1983b) remarked that the

> truth is that I would have loved to play the 'giant killer', as Freeman is now trying to do. (What graduate student wouldn't?) But I couldn't. I had found the village and behavior of its inhabitants to be much as Mead had described. *Coming of Age in Samoa* was like a map that represented the territory so well that I met few surprises when I arrived. (p. 15)

With respect to his points of disagreement with Mead, Holmes (1987) stated the following: "What I did find in an exhaustive review of Mead's writings on Samoa was that the culture was not quite as simple as she would have had us believe. In other words, she often overgeneralized, a common failing among novice fieldworkers" (p. 103).

At the same time, Holmes was quite critical of Freeman's claims about Samoan culture, and he devoted a chapter of his book to a critique of Freeman's refutation. Holmes (1987) said, for example, he "never saw the society as inflexible nor the Samoans as aggressive as Freeman characterizes them. Actually, [he] saw the society as one in which people went to extremes to avoid conflict and to arrive at compromises" (p. 104).

Although Holmes believes that Mead was correct with respect to her coming-of-age thesis, he does not feel that he can support her contentions regarding several other features of Samoan culture and character. These features include her claims regarding the absence of *intense* sexual/romantic relationships and the extreme casualness of Samoans (i.e., that there are no "high stakes," no "fights to the death," or little competition). However, he does not think that these bear upon the issue of coming of age. Holmes believes, as do I, that it was unnecessary for Mead to have made such arguments about Samoan culture and Samoan character to support her coming-of-age thesis. In other words, even if she were wrong on these cultural and characterological matters, she could have been right on the coming-of-age issue.

Finally, it is worth mentioning that Holmes reviewed the extensive personality testing that has been conducted on Samoans. Holmes (1987) concluded that there emerges "a profile of Samoans very much at odds with that put forth by Derek Freeman. [It is] strongly supportive, however, of Margaret Mead's conceptualizations" (p. 136). This profile depicts a low level of creativity, high conformity and benevolence with low independence, and strong tendencies for deference, order, self-abasement, and endurance (pp. 127–136).

Schoeffel and Meleisea

The research team of Penelope Schoeffel and Malama Meleisea conducted several joint and independent studies of Samoan culture (Meleisea, 1987a, 1987b; Meleisea & Meleisea, 1980; Schoeffel, 1979, 1986; Schoeffel & Meleisea, 1983). Based on their academic and personal knowledge of Samoan culture, they produced one of the clearest evaluations of Mead's work in response to Freeman's critique. Although their evaluation does not always bear directly on her coming-of-age thesis, it is worth considering their joint testimony, because it helps us put into context Mead's observations and conclusions concerning Samoan culture and character.[1]

Schoeffel and Meleisea (1983) prefaced their commentary on Mead's work with the assertion that there is a "dualism inherent in Samoan culture" (p. 58). This dualism involves the distinction between "actual individual behavior," as acted out by Samoans in their day-to-day lives, and "ideal social behavior as it would be described by Samoans, particularly adult Samoans of high rank" (p. 59). They argued that although Mead glimpsed this dualism, she did not fully understand it and she tended to confuse the two. They also noted, however, that Freeman has done the same thing. For instance, Freeman's position regarding sexual behavior

---

[1]The following self-description explains why Schoeffel and Meleisea (1983) can be considered experts on the matter:

> Our contribution will be respectively that of an anthropologist who carried out four years of research on the status of women in Samoan social structure and culture, and that of an historian who has devoted the past four years to the formal study of twentieth-century changes in Samoan lands and titles and also "came of age" in Samoa between 1948 and 1972 in a village where his ancestors had lived for the preceding seven generations. (p. 58)

is argued at the level of *ideal* behavior, whereas Mead's is at the level of *actual* behavior. Schoeffel and Meleisea (1983) explained this as follows:

> Freeman disputes the testimony of Mead's informants on the grounds that sexual topics are not freely discussed by Samoans and that her informants deliberately duped her with their accounts. Schoeffel's field notes suggest [1979], to the contrary, that Mead was not duped and that clandestine love affairs, not dissimilar to those related by Mead, are not in the least uncommon. But the crucial point is that they are clandestine and, as Freeman points out, severely punished if they become publicly known. The general Samoan attitude is that, without careful surveillance, adolescent girls and boys will engage in illicit sexual relations. Female virginity is an aristocratic value and, in so far as all Samoans like to consider themselves of aristocratic descent (regardless of their actual rank), family prestige can be enhanced or damaged by public evaluation of the chastity of its unmarried family members. (pp. 59–60)

They continued by saying that "Mead's contention that there was 'passive acceptance' by Church authorities of premarital sex accords well with [their] own observations" (p. 62). Thus, Schoeffel and Meleisea found Mead's observations regarding the *incidence* of sexual behavior to be quite plausible and they have witnessed similar behaviors themselves. It appears that Mead simply did not understand sufficiently the dualism of the culture. This issue is discussed again in chapter 6.

In addition to corroborating Mead's observations regarding the incidence of premarital sexual behavior, Schoeffel and Meleisea stated that "[t]here are many other instances of Mead giving a fairly accurate description of common practice which does not accord to formal ideology" (p. 62). For instance, Freeman argued against Mead by declaring that a voluntary change of residence was *not* possible for Samoan children or young people (Freeman, 1983, pp. 204–205; cf. Mead, 1928, p. 40 ff.). On this matter, Schoeffel and Meleisea (1983) stated the following:

> We have several case studies of this: Malama Meleisea once left home and went to another family (to whom he was related by structural definition rather than by close blood ties). He did this when he had shamed and angered his parents in consequence of defying the village pastor. He went to another family until his father's anger had cooled and his father fetched him home a few weeks later. Malama's foster brother attached himself to the Meleisea family, of his own choice, when he was in his early teens. The arrangement gradually became permanent over the years and eventually Malama's father formally adopted the young man so that he could use the family name as his own. Malama's personal experience accorded well with the many biographical accounts collected by Penelope Schoeffel from Samoan informants. (p. 63)

Although the practice has been commonplace until quite recently,[2] Schoeffel and Meleisea noted that it "is also perfectly true that in terms of formal ideology Samoan children are not supposed to disobey their parents and that the transfer of children between households should be a formal transaction between or among the adult members of the family" (p. 63).

This testimony provided by Schoeffel and Meleisea assures us that Mead's book is not "all rubbish" as some would have it. Moreover, it supports the observations presented earlier in this book that Mead and Freeman viewed Samoan culture from different perspectives. Although Schoeffel and Meleisea are particularly critical of Freeman for not acknowledging the dualism of Samoan culture, and despite their general support for Mead's observations, they do criticize her work in a way that helps to identify areas where she might have erred. In reference to the "dualism" they believe that Mead overlooked, they wrote:

> The essential weakness of Mead's analysis is her confusion or failure to distinguish between actual and ideal categories of behavior. The actual sexual conduct of adolescent girls was confused with ideal rules about how adolescent girls ought to behave. But when it came to their emotions, she reversed the error. In her contention that there was an absence of poignant choices, deep emotions and conflict in the lives of Samoan adolescents, she was echoing the prescriptive, formal, public demand that they should not *display* strong emotion or conflict. Mead's own abundant data confirms the existence of gloomy or passionate emotion and violent conflicts in adolescent girls. Her classification of the cases as 'deviant' conforms neatly to the *Samoan* ideal that people, particularly young people, ought not to display evidence of emotion; that such feelings ought to be contained, repressed or denied. (p. 67)

All told, the expert testimony does two things: It corroborates Mead's coming-of-age thesis (Holmes), and it helps to clarify the criticisms raised by Freeman concerning Mead's account of Samoan behavior and ethos (Schoeffel and Meleisea). Another source of evidence is available that helps us to arrive at a verdict in this case. This evidence is obtained with a return to the scene of the alleged crime and with an examination of the geographical and historical factors relevant to the controversy.

---

[2]It is now less economically viable for families to take in young people. This trend will increase because of the introduction of education (and associated school fees and expenses) and the changing nature of childhood in a wage economy. Thus, as in other countries affected by the West, Samoan children and young people are changing from being economic assets for families to being economic liabilities. This is the case because their decreased contribution of domestic labor is not proportionately replaced by wage labor, and there are outside expenses associated with raising them.

MATERIAL EVIDENCE

Geography and History

Freeman has steadfastly maintained his position that Mead's observations on the island of Ta'u in the 1920s can be evaluated with evidence from other Samoan islands collected several decades later. The following passage appears to represent his position:

> I do *not* believe ... that the communities studied by Mead and myself have ever been exactly "the same." ... What I do consider, however, is that as historically related communities with a common language and a common culture, they are, for a wide range of behavioral and cultural phenomena, sufficiently similar to make comparisons between them *in respect of these phenomena*, rationally justified and empirically significant. (Freeman, 1985, pp. 912–913)

Many commentators on the controversy find it difficult to accept such a claim, however. For example, Schoeffel and Meleisea (1983) remarked that they "have the greatest difficulty in believing that there have not been major changes in Samoan society and behavior from one decade to another, let alone over the past fifty-five years since Mead commenced her study" (p. 63). Certainly, my training as a sociologist leads me to have great difficulty in believing that there would be no differences in behavior, particularly deviant behavior, on an island like Ta'u (Mead's reference point) in comparison with an island like Upolu (Freeman's reference point), in spite of a shared culture (cf. Turnbull, 1983).

Let us begin, then, with a look at the geographical differences between the two reference points that might account at least in part for the contradictory conclusions drawn by Mead and Freeman. In chapter 2, two maps of the Samoa archipelago were provided. These are meant to help readers understand how the geography of the Samoan islands might have affected the *possibilities* and *limitations* of Samoan culture and behavior over the centuries.

Referring to Fig. 2.1, it is immediately obvious that Ta'u is much smaller than Upolu. In fact, Ta'u is only 28.5 square kilometers (Hunt & Kirch, 1988, p. 156),[3] whereas Upolu is 1,115 square kilometers. Thus, Upolu is over 30 times larger than Ta'u. Moreover, Ta'u has proportionately more mountain slope in relation to arable and habitable land. With its highest point at 965 meters and only 41% of its area having less than a 30% slope, a mere 12 square kilometers is usable. With respect to

---

[3]Other sources estimate its size to be 39 square kilometers (e.g., Swaney, 1990).

population, in the 1960s Upolu had a population of 90,000 living in over 200 villages, compared with Ta'u in the 1920s comprising 4 villages with a population of about 1,200.[4] Thus, the population of Freeman's reference point is some 75 times greater than that of Mead's reference point.

The ruggedness of the islands in which Mead carried out her research is described by Hunt and Kirch (1988) in the following manner:

> The Manu'a Islands of Ofu, Olesega, and Ta'u form a separate cluster at the eastern end of the Samoan archipelago. Mutually intervisible, they are separated from Tutuila to the west by 100 km of often turbulent ocean which reduced the frequency of voyaging contacts with the larger islands. . . . Ofu, Olesega, and Ta'u are remarkable in their *dramatic topography; steep-sided, majestic volcanic cones thrust out of the turbulent waters, with summits often shrouded in clouds* . . . The smallest of the principal Samoan islands, *their steep topography offers little area suitable for settlements and gardens* . . . *Coastlines are rock-bound, with narrow fringing reefs only in places*, restricting the possibilities of marine subsistence. (p. 155, italics added)

In contrast, Upolu, the island upon which most of Freeman's research was carried out, "extends about 72 km from east to west and up to 24 km from north to south . . . [and it] . . . has a chain of volcanic peaks running from one end of the island to the other, with hills and *coastal plains on either side*" (Douglas & Douglas, 1986, pp. 510–511, italics added). Thus, although rugged in areas, Upolu has far more area suitable for habitation. The ruggedness of Ta'u can be seen with the contour lines in Fig. 2.2, as can that of Ofu and Olesega (the three islands constitute all of Manu'a).[5]

---

[4]Ta'u's population was about the same in 1980—Hunt and Kirch (1988) put it at 1,146. An estimate from 1854 placed the entire population of Manu'a at 1,275 versus 15,587 for Upolu (Ward, 1967, p. 365). Another source (Park, 1980) traced the population pattern of Manu'a over this century, specifying a range of 1,756 in 1900 to a peak of 2,800 in 1950. For 1926, Park estimated it at 2,060.

It should be noted that population estimates for Samoa are generally considered rather unreliable, and that many estimates are likely inflated through the inclusion of people who have moved but who are still considered family members (e.g., they may actually be residing on Tutuila or have emigrated to Hawaii or San Diego). Hunt and Kirch (1988, pp. 155–156) argued that for all of Manu'a "the resources of the land and sea are sufficient to support a modern population of 1,700 persons and it is certain that the late prehistoric population was several times greater than this figure." Thus, the most liberal estimate of the population that all of Manu'a could ever sustain would be about 5,000 people. Assuming that Ta'u would sustain 60% of Manu'a's potential population of 5,000, Ta'u could still sustain only one thirtieth the population of 1960s Upolu.

[5]According to Hunt and Kirch (1988), Ofu and Olesega are 3.4 and 4.5 square kilometers, respectively; their highest points are 638 and 494 meters, respectively; their areas with less than a 30% slope are 9% and 10%, respectively; and in 1980 their respective populations were 254 and 340.

MATERIAL EVIDENCE 55

Two photographs taken during my visit in February 1991, bring these maps to life. Figures 3.1 and 3.2 show much of Mead's research site. Although the buildings have changed many times since as a result of hurricane damage, the size of the villages has not. Figure 3.1 captures about one half of the village of Luma, the village in which Mead lived while she did her research. Note how small this village is, being squeezed between the ocean and a marsh (the unused area in the left of the photograph is marshland). Note, as well, the proximity of the houses, which affords little privacy. Figure 3.2 provides a view of about half of the village of Faleasao, situated over a hill from Luma. Observe, as well, its small size and the proximity of all of the dwellings. Note, also, that there is no room for this village to grow—it is situated on a beach immediately below the cliffs of an old volcano cone. Together these two figures depict about one half the area of the villages in which Mead's informants lived. Such a tightly knit setting hardly looks like the "hotbed" of delinquency and psychological malaise claimed by Freeman; on the contrary, this is the type of setting in which one would expect a high level of community control over behavior. It is hard to imagine how such small communities would have survived over the centuries if disruptive deviance were not somehow minimized.

Obviously, this topographical and photographic evidence does not constitute "proof" of anything, but it does give the reader a frame of reference

FIG. 3.1. Luma Village.

FIG. 3.2. Faleosao Village.

with which to evaluate other claims. Accordingly, with this knowledge of Ta'u, we can now more easily consider other information about Ta'u that highlights its uniqueness in comparison with the rest of Samoa, particularly during Mead's visit. These unique circumstances, all of which cast doubt on the reasonableness of Freeman's critique, include the following historical and political factors.

American Samoa had not experienced any wars for several decades, and it had been annexed by the United States at the turn of the century. Manu'a, of which Ta'u is a part, had recently seen its long-reigning line of royalty end, so an authority void would have existed. In addition, Mead noted repeatedly how Christianization, still in progress at the time, had tempered many harsh practices found in the traditional culture as well as ameliorating intervillage and interisland hostility. Thus, the period before, and during, her visit seems to have been one of political harmony.

After presenting a similar analysis, and concluding that "things were pretty quiet while Mead was there," Schoeffel and Meleisea (1983) acknowledged that they "have some sympathy with the problems of those who carry out brief research in quiet times" (p. 64). They state this because Schoeffel studied a village between 1976 and 1982 that had been investigated by B. Lockwood (1971) 10 years earlier, and

[s]he was amazed at the difference between her own observations and those of Lockwood; apparently in the space of ten years a progressive, prosperous, well-governed and peaceful community had become a strife-torn, faction-ridden, economically stagnant village in which leadership was virtually absent. (p. 64)

The differences observed by two investigators in the same village only 10 years apart should make us wary of claims such as Freeman's that radical differences in behavior are not to be expected in different parts of Samoa, especially at different points in time. Schoeffel and Meleisea accounted for the differences in the following manner:

While Schoeffel's first impression was that Lockwood must have mixed up his villages or his field notes, Meleisea's longer experience with the events of the past and those of Schoeffel's particular ethnographic present, led [them] both to realize that the two highest ranking chiefs of the village had been in the prime of their political careers, both nationally and locally, when Lockwood conducted his inquiries. In 1978 they were aged and ill, and the village was in the grip of a series of successive rivalries among a collection of aspiring chiefs and would-be chiefs which had disrupted the economic prosperity and social tranquillity of the village. (pp. 64–65)

The geographical and historical analysis given here, along with earlier analyses of Samoan culture, make it clear that, as in other cultures, Samoan behavior can take many forms. Like other cultures, therefore, Samoan culture can manifest itself in a myriad of ways, depending on the forces influencing it at a particular time and place. But this should come as no surprise for those who are not operating at the level of stereotypes. In the context of Schoeffel and Meleisea's just mentioned experiences, an observer might find "Freeman's Samoa" in one village, whereas another observer might find "Mead's Samoa" in the next; neither observer would be wrong, and both would have seen different facets of the "same Samoa." Indeed, as Meleisea (1987a, p. 31) noted, Samoan culture is a "living culture" in which "nothing is ever fixed; small changes and adjustments are constantly taking place so that the actual situation often differs from the ideal model which people may have of society and its rules."

Another way to evaluate Freeman's claim is for readers to consider how conditions have changed in their own culture since the 1830s—when the missionaries first arrived in Samoa; or since the 1920s—when Mead did her research. For example, consider how much life has changed since 1920 for those "coming of age" in California or Colorado, in Ontario or Quebec, in New South Wales or Victoria, or in England or

Scotland. Consider, as well, how sexual mores and childrearing practices have changed. Sexual mores and childrearing practices are aspects of the same culture but it is unreasonable to assume there have not been significant differences over such spans of time and place.

Three very important conclusions can be drawn from the evidence just presented with respect to the Mead–Freeman controversy. First, given the actual and potential uniqueness of Ta'u, there is sufficient reason to doubt Mead's blanket generalizations from Ta'u to the rest of Samoa, as well as Freeman's contention that coming of age on Ta'u in 1925 would have not been significantly different from doing so in Western Samoa in the 1960s. The second conclusion pertains to Freeman's claim that Samoan culture is essentially homogeneous and can be understood from one research site at one point in time. His position on this matter is contradicted by the research of other scholars of Samoan culture. The third conclusion is that if Mead had initially titled the book something like *Coming of Age on Ta'u*, and not generalized to other Samoan islands, there would have been little basis for a critique. In her scholarly work, Mead did restrict her generalizations to Manu'a (e.g., 1931, p. 545) and the companion book to *Coming of Age* specifically restricts itself to Manu'a—indeed, it is called *The Social Organization of Manu'a*. However, these do not nullify the inappropriate generalizations made in *Coming of Age*.

Mead's Own Defense

Minor criticisms of Mead's work had been circulating throughout the anthropological community for some time before Freeman's book was published (e.g., Shore, 1982; Stanner, 1953). But criticism is supposed to be "healthy" in science—evidence that scientists are doing their job of scrutinizing each other's work. Although Mead responded to these criticisms on several occasions, she apparently did not feel that any major revisions of her book were necessary, and she never provided a full defense of her position. She did, however, provide two partial responses to criticisms of her work. The first is in the introduction and conclusion to the 1969 edition of *Social Organization*, and the second is in the Preface of the 1973 printing of *Coming of Age*.

In the latter source, Mead commented on encounters she had with Samoan university students who were embarrassed by her frank and sometimes uncomplimentary portrayal of their grandparents (cf. F. Wendt's, 1984, criticisms of the book). She lamented that they and other readers tended to disregard the fact that the study was carried out in the 1920s and describes life then. In her own defense, she stated: "It seems more than ever necessary to stress, as loud as I can, this is about the Samoa . . .

MATERIAL EVIDENCE 59

of 1926–1928. When you read it remember this. Do not confuse yourselves and the Samoan people by expecting to find life in the Manu'an islands of American Samoa as I found it."

She continued in that Preface to note that she did not want to alter the contents of the book (despite its many reprintings) because it was a historical document and she did not want to change its style of writing. In her words:

> I can emphasize that this was the first piece of anthropological fieldwork that was written without the paraphernalia of scholarship designed to mystify the lay reader and confound one's colleagues. . . . *I did not write it as a popular book*, but only with the hope that it would be intelligible to those who might make the best use of its theme . . . (italics added)

Despite her protestations, however, Mead was not being entirely forthright. The book has been widely read by the general public; in fact, it is "the most widely read anthropological book ever published [with] sixteen translations and millions of copies sold" (Muuss, 1988, p. 139). Moreover, it would appear that she wrote in a way to make it marketable, particularly with the inclusion of the academically controversial chapter entitled "A Day in Samoa." This is the second chapter of the book and was likely placed near the beginning so as to grab readers' interest before they encountered the more mundane details of her account. In this chapter, she employed the now familiar popularizing technique of providing a composite sketch of daily life. Unfortunately, to imply that all of those events could happen in one day is misleading (cf. Holmes, 1983b) and not ethnographically acceptable. On the other hand, Mead instructed that "[f]or the scholarly reader, there is a new edition of *The Social Organization of Manu'a* . . . revised in the light of contemporary ethnographic theory" (Preface, 1973 printing of *Coming of Age*).

Yet, contrary to these declarations, it seems that Mead was occasionally embarrassed about *Coming of Age* because of its nonacademic style. For example, in the introduction to *Social Organization* (1969) she noted that "I incorporated in [*Coming of Age*] a section originally intended for this monograph, called 'A Day in Samoa,' which I had decided was too literary in character for the style of a Bishop Museum monograph!" (p. xvii). And, when it was subjected to academic scrutiny by Lowell Holmes, Mead equivocated in terms of how authoritative she saw the book to be:

> Holmes' quotations are from reprints of paperbacks which make the whole dating system into pulp. I wrote *Social Organization of Manu'a* (1930) after I wrote *Coming of Age in Samoa* (1928), and the former should therefore be regarded as the more complete and definitive. (p. 224)

In *Social Organization*, Mead went on to comment on what was then mounting criticism of her characterization of Samoan "character" as one of "mildness and low affect" (p. 226). She raised two possibilities regarding the discrepancy between her own analyses and those of others. The first possibility was that the Ta'u that she witnessed in the 1920s was a historical anomaly in which there was a "temporary felicitous relaxation of the quarrels and rivalries, the sensitivity to slights and insults, and the use of girls as pawns in male rivalries" (p. 228). The second was that she saw the culture through the eyes of her informants—young females. In Mead's (1930/1969) words:

> The other possibility is that to the young girl, herself either a virgin but not a *taupou*, or experimenting quietly with lovers of her own choosing, uninvolved in the rivalries that were related to rank and prestige, moving gently, unhurriedly into adulthood, the preoccupations of the whole society may have seemed more remote than they would have appeared from any other vantage point. *And this is the vantage point from which I saw it*. I was alone, very slight and smaller than the Samoan adolescent girls. My subject of research called for my spending the largest proportion of my time with them. . . . My primary task was to get to know and understand adolescent girls; the ethnography of this monograph was a by-product, an extra dividend. (p. 228, italics added)

Those who now stand in judgment of Mead's work have a responsibility to acknowledge her admission regarding the possible discrepancies noted here regarding Samoan character and to appraise her explanation for them. Her admission also requires, however, that the generalizability of some of her findings beyond Ta'u of the 1920s and beyond the experiences of those "coming of age" be viewed with caution. In chapters 4 and 5, I act on this caution and carefully examine the social history of adolescence in Samoa since early missionary contact.

Finally, Freeman and some of his supporters make much of the fact that Mead did not live in a Samoan household. In reality, she lived in a medical dispensary in the village of Luma (see Fig. 3.1). In one of her letters written shortly after her arrival, Mead (1977) said that the "people in these villages are right at my very doorstep" (p. 36). This lodging gave her an excellent *neutral* site from which to study all of the inhabitants of the villages, not just those who happened to be members of a household in which she might have stayed. Had she stayed in a Samoan household, she would have likely been constrained in her movements to other households for fear of offending her hosts (cf. Swaney, 1990), and, as she said, she might have been caught up in the various intra- and inter-*aiga* (family) rivalries that characterized village living. As she wrote in the acknowledgment section of *Coming of Age*, the dispensary furnished her "with an absolutely essential neutral base from which [she] could study all the individuals in the village and yet remain aloof from native feuds and lines of demarcation." Thus, she managed to find what she described as "an excellent arrangement."[6] Here is how Mead (1977) described her quarters—note how it allowed her to interact with her young informants on their terms:

> My room is half of the back porch of the dispensary quarters building. There is a loosely woven bamboo screen which divides my room from the porch outside the dispensary and here the Samoan children gather to peek through the holes and display a few English words or chatter endlessly in Samoa about Makelita's various belongings. . . . At night I push back the curtain which divides my room off at the other end, put away the chairs, push back the tables and there is plenty of room for a small *sivasiva*—dance. The young people bring their guitars and ukuleles and dance for me. A few new ones come every night and it gives me an excellent opportunity to gradually learn their names . . . (p. 37)

A number of observers condone Mead's decision to carry out her research in this way. Turnbull (1983), for example, argued that

> on such a small island, to have settled in with one family in one village would have placed her in special and perhaps restricted relationships with other families and other villages. Further, given the open architecture of the traditional Samoan house (a roof with neither outside walls nor inner partitions), no privacy would have been possible; and for her inquiries into the delicate topic of the adolescence of girls, Margaret Mead judged correctly that both she and the girls would require privacy. She knew perfectly well what limitations this imposed. Even today a comparable choice would by no means be uncommon, and better by far than any pretence of "going native." (p. 33)

In the final analysis, the issue of Mead's choice of residence is really a red herring with respect to her coming-of-age thesis. The fact is that her primary concern was not an ethnography of village life. Rather, her primary concern was with the experiences of the young females in three villages. From her base of operations, she was able to interact with and interview these informants, without being distracted by the "domestic politics" of any one Samoan household.

---

[6]Dr. Lee Cranberg, who worked as a physician in the village of Luma during 1979–1980, found village life there to be "transparent" and easy to integrate into. He had little doubt that Mead could have developed a clear picture of village life (personal communication, November 27, 1991). In his words, "[t]hat she shared accommodations in the midst of the village with an expatriate American couple would not have prevented her from sharing her day-to-day life with Samoan neighbours" (Cranberg, 1983b, p. 19).

A VERDICT ON MEAD'S COMING-OF-AGE THESIS

Before readers deliver a verdict based on the evidence presented here, it is appropriate to deliver several charges to them as the "jury."

First, we cannot go back in time to the Ta'u of 1925–1926, a time during which Mead was the only record keeper there. Indeed, it is fruitless to try to assess her thesis today with ethnographic or quantitative social/psychological measurements because life on Ta'u has changed significantly (cf. Holmes & Holmes, 1992). Therefore, in evaluating Mead's claims about that time and place we must rely on inference and plausibility, as well as on corroboration from other observers, particularly those who were closest in time and place to her research site.

Second, it is unrealistic to expect her study to be flawless, especially given the pioneering nature of her research. The recognition that some of her observations cannot be verified, or that some of her conclusions are tenuous, certainly does not justify dismissing the entire book (cf. Appell, 1984, p. 205). It is this type of unreasonableness that has generated more "heat than light" (Feinberg, 1988, p. 656) in this controversy. Thus, with very little prior research to go on, it is understandable that she might have overlooked some matters and overemphasized others. What many of her critics seem to have forgotten is that this was one of the earliest studies of adolescence and she therefore had very little prior biological or social research on which to base her work. In short, it is easy to sit in harsh judgment a half century later and ignore the fact that she blazed a trail that has led to a body of knowledge that can now be used to judge her work.

Third, we must keep in mind that even if Mead was wrong in some of her characterizations of Samoan culture, character, and ethos, she still may have been right in her conclusions about coming of age. After all, the study was subtitled *A Psychological Study of Primitive Youth for Western Civilization*, so it was primarily a psychological study of adolescence, not an ethnography of Samoan culture. Moreover, she may have been right about the ease of adolescence but she may have been wrong in her reasons for arriving at this conclusion.

Finally, it must be remembered that what is at issue here is ethnographic-type research, a research technique that can be high in internal validity (accuracy), but low in external validity (generalizability). As Bernard (1988) noted, two of the principal weaknesses of ethnographic research are that "(a) it is difficult for other researchers to replicate an ethnographer's findings . . . ; [and] (b) whatever an ethnographer learns about one village or island may have little to do with other villages or islands in the same general cultural region" (p. 146). Mead's research is vulnerable to both weaknesses, but it is hardly appropriate to hold her personally responsible for the shortcomings of a research method.

With these qualifications in mind, and after weighing all of the evidence presented thus far, I believe a reasonable verdict lies in favor of Mead's coming-of-age thesis. In these inquiries, little reason could be found to doubt that Mead reported accurately what she saw and was told during her study; to suspect "fudging" on her part; or to assume "duping" by her informants. Moreover, Freeman's claims regarding a storm and stress ostensibly characterizing Samoan adolescent behavior on Ta'u in the 1920s are weak and the evidence he presents can be interpreted in ways that do not contradict Mead's conclusion. Finally, not only are Freeman's arguments regarding adolescence easily refuted themselves, there is also sufficient corroboration of Mead's conclusion from other sources. More corroborative evidence is examined throughout the remainder of this book.

Although there is little in Freeman's critique that constitutes an actual scientific refutation, it nevertheless appears that Mead compromised several canons of social science by sensationalizing parts of her account to make her book more marketable. In addition, she appears to have capitalized on romantic notions about the South Seas and sexual stereotypes about Polynesians, although not nearly to the extent claimed by some of her critics. We have also seen evidence that she engaged in some unsubstantiated speculation regarding why "coming of age" there was one of ease, and she had insufficient grounds for assuming that her observations and conclusions could be generalized to the rest of Samoa. Thus, although her main thesis appears to be supported by the evidence, it seems that she did not exercise what is now considered the appropriate scholarly restraint expected of an academic in presenting and interpreting findings. On the other hand, she said that she wrote the book for "teachers, parents, and soon-to-be parents" (p. 1973 edition), and she was ostensibly trying to convey a message to that audience.

It is unfortunate that the controversy has focused on a few notions about Mead's book, thereby overlooking what is otherwise a "rich and sensitive" account (Feinberg, 1988, p. 656). Even with its faults and limitations, if it is read as a semi-popular, pioneering study with cross-cultural implications, it can be appreciated as a valuable historical document and a landmark study. As such, it advanced the understanding of adolescence in its time, and anticipated later developments in the field (such as the concept of the moratorium period and the socioeconomic forces prolonging youth and adolescence). Chapter 6 reconstructs her portrait of how Samoans would have come of age at the time. From this reconstruction we get a much better idea of how rich and sensitive her work is.

In view of a verdict in favor of Mead's thesis, what do we have to say to Freeman about his years of work on these issues? First, his critique has alerted us to the limitations of Mead's research, and potential problems with its generalizability. For this we are grateful. And, second, he has inadvertently sounded the alarm of an unfolding tragedy in modern Samoa; namely, the difficulties facing many young Samoans because of the cultural disenfranchisement brought on by Western influences. For this we are also grateful.

Were it not for this latter implication of the controversy, we could reasonably end this book here. But, in chapter 7, I take up the issue of the cultural disenfranchisement of Samoan youth, after constructing a social history of adolescence in Samoa. This allows us to better understand what has happened as a result of contact with the West. At the same time, it gives us the opportunity to assess the historical veracity of Mead's work. Completing this task allows us to draw a conclusion as to whether Mead "fabricated" a Samoan culture, out of touch with the rest of the culture, as some have charged. As we see, these two issues actually dovetail because, in assessing the historical accuracy of Mead's research, we come to understand how drastically "coming of age in Samoa" has changed. Thus, it is within the context of Samoan history that we come to fully understand the Mead–Freeman controversy.

CHAPTER FOUR

A Social History of Adolescence in Samoa: Precontact Culture

Freeman contended that Samoan culture is, and has been, essentially homogeneous throughout the archipelago and that very little change relevant to the controversy took place between the time of Mead's study and his own research. In a similar vein, he insisted that adolescence has been stressful in Samoa since before Mead's study and before contact with the West. The evaluation presented here casts Freeman's case into serious doubt. But, that appraisal was based largely on evidence of a more *inferential* nature, so it could be argued that the review of Freeman's case is not complete. In order to more decisively judge this aspect of the controversy, it is appropriate to examine evidence of a more *descriptive*, historical nature that allows us to draw conclusions regarding social change and stability in Samoa (chapters 4 and 5) and the plausibility of "Mead's Samoa" (chapter 6). The purpose in carrying out this comparison is to determine whether Mead's Samoa fits with the "trajectory" of social change set in motion by contact with the West and the social engineering undertaken by the missionaries.

In compiling this historical evidence it became clear to me that not only does Mead's Samoa fit the pattern of change, but that the direction of this change has continued to the present. It also becomes clear that Freeman's assumption of an absence of significant social change simply begs the very important sociological question of the *nature* of the changes affecting this region. Indeed, as we see, the following sociological investigation of the history of adolescence in Samoa reveals a fascinating yet depressing saga of the demise of a well-structured and meaningful adolescence before contact with the West, and the creation of a conflictual and alienating adolescence after a century and a half of contact with the West. By denying the relevance of social change in Samoa for the Mead–Freeman controversy, Freeman overlooked the possibility that the turmoil he witnessed among Samoan youth in the 1960s was a product of this change. As is seen later, the culmination of this change has been disastrous for many of those now coming of age there.

Many written sources have been consulted in reconstructing this social history of Samoan adolescence: the accounts of missionaries, written mainly in the 1800s; the results of studies conducted by anthropologists during the 1900s; fictionalized writings and poetry representing the Samoan "spirit"; and, contemporary documents including government reports, newspaper articles, and documentary films. Much of this evidence is presented verbatim from these sources. Consequently, some readers may find the long quotations tedious to read. Given the controversial nature of this topic, however, it is appropriate to reduce my synthesis of this material, and to let readers judge it for themselves.

COMING OF AGE IN PRECONTACT SAMOA

The available written documentation reveals a number of features of precontact Samoa that should have provided for a relatively continuous and stress-free passage from childhood to adulthood.[1] Perhaps most crucial to the issue of continuity was the fact that Samoan children were not protected from the hardships and "evils" of the world. In contrast to the child-centered conventions of contemporary Western societies, Samoan children were not indulged in a protected fantasy world that left them wholly dependent on adult supervision and care; they were not an economic liability to the household, rather they were an economic *asset* and investment; and they lived in the real world of adults, witnessing most of what adults witnessed. In short, by their teen years Samoan young people were already "worldly" individuals in their own right, with a working knowledge of that world, and they possessed socially recognized capacities and skills upon which to build an adult sense of identity (cf. Holmes, 1987). In short, there was a well-established *quid pro quo* between the young and adults that dictated that all individuals contributed

---

[1]I am purposely excluding Mead's writings from most of the present chapter, because one of the objectives of this chapter is to assess the plausibility of Mead's account of Samoan adolescence by estimating what conditions were like for those coming of age in precontact society. As we see, Mead's research provides the first *thorough* record of Samoan childhood and youth before the Samoan life cycle was drastically altered by Western influences. As we also see, Mead's *Coming of Age* is important because she made reference to many of the things that had been, or were being, altered by the missionaries.

to the collective to the extent that they were capable, and all received their share as they needed. There is little of a *quid pro quo* between adults and children in Western societies, and it is being lost in contemporary Samoa, as we see later in this book.

Those familiar with the topic will know that most "primary" cultures expose their young to the "real-world" conditions of the adult culture. But Samoan culture had the added advantage of a relatively fruitful natural environment. Thus, although young Samoans contributed labor to their family and community, the amount of labor was minimized for all Samoans by the benign climate and lush environment. This allowed Samoan culture to give those coming of age the time to slowly learn the tasks of the culture, and to gradually increase their labor contribution as they entered adulthood. Such a "moratorium," as it is now called, can play a crucial role in alleviating the stresses of life that might otherwise accompany major life transitions (cf. Erikson, 1968; and chapter 8, this volume).

On the matter of this natural "affluence," Holmes (1987) estimated that on Ta'u an average family could "provide itself with ample food by working its plantations only about three days a week" (p. 34). He also noted that "[w]ithin the memory of the Manu'an people there is only one occasion, a hurricane, when there was a shortage of food. . . . In Samoa no one goes hungry. The extended family provides for all its members" (p. 34). The labor of children, however minor the tasks involved, made it possible for all members of the community to enjoy the benefits of this benign environment. This is corroborated by Greksa, Pelletier, and Gage (1986) who, after reviewing the relevant literature, concluded that "productive males in precontact Samoa probably needed to expend no more than 6–12 hours of work per week in agricultural labor. Even if one assumes a highly conservative estimate of 15–20 hours of agricultural labor per productive male per week, this is not a great deal of work" (p. 303).

Meleisea (1987a) presented a similar portrait of life in precontact Samoa. In his words:

> Everything we know about Samoa in the early 19th century indicates that the Samoan people had achieved a very abundant comfortable way of life in which everyone was well fed and well housed. Things looked so good to European seamen in the 19th century that many of them ran away from their ships to live with the Samoans. T.H. Hood . . . remarked: "on the whole, a happier race of people could not be found than the Samoans. A scowling and discontented face is seldom seen; want or poverty is unknown, and nature has showered upon their country her choice gifts." (p. 26)

The inclusion of such descriptions is not intended to romanticize or idealize precontact Samoan culture, but to give readers some idea of the type of *environment* in which young Samoans would have come of age. The point of including these descriptions is to outline the felicitous environment that made it possible for certain social institutions to arise that were not oriented directly to group subsistence. Let us now examine these institutions.

We know much more about the way males came of age in precontact Samoan society than we do about the coming of age of females. For males, the principal rite of passage on the path to full adulthood occurred at age 16. The initiation entailed experiencing the long and painful process of body tattooing, from the waist to below the knees. The missionary George Turner (1861/1986) wrote of the importance of the tattoo for the adult male social identity:

> the man who was not tattooed . . . was not respected. . . . Until a young man was tattooed, he was considered in his minority. He could not think of marriage, and he was constantly exposed to taunts and ridicule, as being poor and of low birth, and having no right to speak in the society of men. But as soon as he was tattooed, he passed into his majority, and considered himself entitled to the respect and privileges of mature years. When a youth, therefore, reached the age of sixteen, he and his friends were all anxiety that they should be tattooed. . . . On these occasions, six or a dozen young men would be tattooed at one time. . . . In two or three months the whole is completed. The friends of the young men are all the while in attendance with food. (pp. 87–89)

Marquardt (1899/1984) reported at the turn of the century that males who were not tattooed were considered cowards and were highly stigmatized:

> Especially the women despised them and the chiefs refused to accept food from their hands which they called stinking. The contempt for men who lack the national adornment has been weakened little by the influence of the missionaries but nevertheless continues until this day. An untattooed man is still not popular and many a father refuses him his daughter's hand. It is no rarity that young people who are being prepared for the mission service turn their backs on the mission schools for the only reason that the "students" are forbidden to get tattooed. (p. 8)[2]

When this tattooing was complete, the male became a member of the *'aumaga*, the group of "untitled men" in the village. The *'aumaga* was

---

[2]Grattan's (1948) record supports Turner's and Marquardt's observations: "In earlier custom an untattooed young man was not regarded very seriously either by his fellows or the girls of the village" (p. 112).

responsible for accomplishing work projects and daily maintenance of the village. Through hard and skillful work and service (*tautua*), the untitled man could earn respect and reputation in the community, and perhaps eventually gain greater status by being elected a *matai* (chief) by his *'aiga*. Each *'aiga* had at least one *matai* who headed their branch of the family and represented them at the *fono* (village council of chiefs).

Young women had a corresponding institution that structured their adolescence—the *aualuma*. Meleisea (1987b) described the importance of the economic and social role played by the young women of the *aualuma*:

> The *aualuma* represented the honour of the *nu'u* [village], for example when a *nu'u* or a district went to war, a high ranking maiden would march at the head of the party. Also, when *malaga* (visiting parties) came to the *nu'u*, the *aualuma* would decorate the guest houses of the *nu'u* and see to the reception and entertainment of the guests. Whereas household goods such as tapa, sleeping mats and thatch pieces were made by groups of wives (who were regarded as being comparable, although somewhat lower status to the *'aumaga*), objects of great value such as *'ie toga* (fine mat) and *'ie sina* (a cloth with a shaggy surface woven from bleached fibers) were made by the *aualuma*. The titular leader of the *aualuma* was the *sa'o tama'ita'i*, a girl or woman from the family of the highest ranking *ali'i* who held a title derived from the name of an important ancestress. (p. 7)

Meleisea also described the correspondence of the positions of young males and females in precontact Samoan culture:

> Work outside the immediate sphere of the *'aiga* was carried out by corporate groups: the *'aumaga*, to which the untitled men belonged; and the *aualuma*, to which all the girls and women of the *nu'u* (as distinct from in-marrying wives) belonged. Each group had its own dwelling houses in the *nu'u* and both were internally stratified according to the rank and status of the *matai* titles of one's *'aiga*. These groups reflected the division of labor and status between brothers and sisters in *aiga*. The *'aumaga* were the *malosi o le nu'u* ("the strength of the village"), its army, fishermen, horticulturalists, cooks and sportsmen. (p. 7)

Of great interest in itself, but also in the context of the Mead–Freeman controversy, is the fact that females also underwent a tattooing ritual. Moreover, knowledge of this practice has been all but lost, due no doubt to the influence of the missionaries and a desire to suppress knowledge of the "heathen" past. Fortunately, a monograph written in 1899 has preserved knowledge of the practice. According to Marquardt (1899/1984), the tattooing of women was still widespread at the time in German-administered Western Samoa. He estimated that about 60–70%

of women sported these tattoos. Marquardt speculated that it was done because whereas "[m]en had themselves tattooed to please women . . . women [did it] to increase their attractiveness to men" (p. 16).[3] The fact that it was "usually carried out at the time of beginning puberty" could be explained, according to Marquardt, because it was "at this time sexual feelings start to develop. A girl wants to please a man and therefore she asks for the means to achieve this purpose and she obtains them" (p. 17). He also recorded that it was done at the same time as the chief's sons were tattooed and that it was "always carried out by male tattooing masters, usually between the girls' fourteenth and sixteenth year. The complete tattooing of the thighs usually takes five to six days" (p. 22). Figure 4.1 shows the types of frontal designs recorded by Marquardt that would typically be sported by males and females (from Ngan-Woo, 1985, p. 67; see Marquardt, 1899/1984, for detailed drawings and photographs of the various components and designs).

During their teenaged years, young Samoans would learn or hone most of the skills, and perform most of the tasks, considered part of the "common-sense" knowledge of the adult world. In addition, many would adopt a specialization that would constitute a foundation of their adult identity. According to Meleisea (1987a):

> Although every adult Samoan knew how to perform the basic economic tasks appropriate to their sex in order to survive, there were dozens of specializations recorded by early observers. The basic economic tasks for men were agriculture, carpentry, hunting and fishing; and for women, weaving, tapa making and oil making. The specializations were based upon elaborate artistic refinements of these basic activities; for example fishing and pigeon hunting were technically very complex and diverse and associated with a lot of ceremony which turned them into gentlemanly sports instead of mere food-getting activities. The master fisherman or master hunter took charge of these activities and used their specialized knowledge to organize large groups of less skilled people. The same was true of house-building, boat-building, wood-carving, tattooing and many other activities. Among women there were specialists in manufacturing the most valuable kinds of mats, tapa cloth, medicines and oils. (p. 33)

Thus, several specializations were open to both males and females within the village division of labor. Status could be gained equivalent to that of a "master" of these specializations through apprenticeships with an established expert. Turner (1861/1986) provided an account of how the Samoan technique of *fale* building was transmitted from one generation to the next:

---

[3]O'Meara (1990, pp. 75–76) also reported that he was told that it was done to increase sexual attractiveness, but he knew of no recent instances of it being carried out.

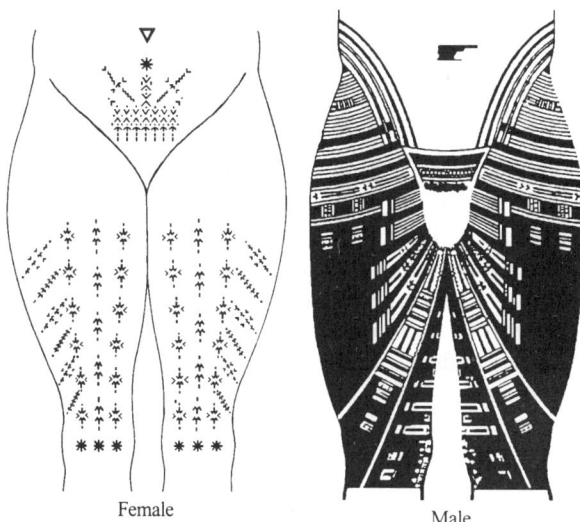

FIG. 4.1. Traditional frontal tatoo designs (from Ngan-Woo, 1985).

> house building . . . is a distinct trade in Samoa; and perhaps, on average, you may find one among every three hundred men who is a master carpenter. Whenever this person goes to work, he has in his train some ten or twelve, who follow him as journeymen, who expect payment from him, and others as apprentices, who are principally interested in learning the trade. When a young man takes a fancy to the trade, he has only to go and attach himself to the staff of some master carpenter, follow him for a few years, until he thinks he can take the lead in building a house himself. (p. 167)

Ceremonies apparently marked the entrance of the young person into the *'aumaga* or *aualuma*, but the records regarding precontact Samoa are not clear. For males, tattooing was definitely a necessary requirement, but it is unclear whether a village feast was held for each male for the actual recognition of membership. Holmes (1987, p. 118) noted, however, that in 1950s Ta'u, the "*matai* accompanies the boy to the village council [*fono*] and presents [a] kava root so that his son will have village recognition when he joins" the *'aumaga*. With respect to entrance into the *aualuma*, the record is also unclear, but Turner (1861/1986) seems to be describing this ceremony in the following passage:[4]

> About the time of entering into womanhood, their parents and other relatives collect a quantity of fine mats and cloth, prepare a feast, and invite

---

[4]The meaning of this ceremony is verified by Mead's (1928) account when she wrote that formerly "[t]he girl's entrance into the *Aualuma* was always, not just occasionally, marked by a feast" (p. 275).

all unmarried women of the settlement. After the feast the property is distributed among them, and they disperse. None but females are present. It is considered mean and a mark of poverty, if a family does not thus observe the occasion. (p. 90)

In precontact Samoa, a man would leave the *'aumaga* only upon being conferred with a *matai* title. Otherwise, he would remain in it for life. Although most males probably aspired to acquire such a title, especially as they approached middle age, some apparently did not experience this ambition (cf. Mageo, 1988). However, Grattan (1948) noted that "progress from untitled to titled rank is the normal aspiration sooner or later of most adult males" (p. 14), and implied that this was the case in precontact culture. Women would leave the *aualuma* upon marriage. We see in the next chapter how this has changed.

ANALYSIS

In sociological terms, it is critical to the continuity and stability of a culture that the young be recruited into adulthood as welcomed and willing participants. The less they are welcomed, the less accepting the young will tend to be of the conventions of the culture awaiting them as adults. Consequently, the less responsible they will feel about willingly passing on the culture as they found it. This was apparently not a problem in precontact Samoa, although it now is in many Western societies, particularly for women. In fact, notwithstanding the changes discussed in chapter 5, in some respects Samoan culture is still considered the most stable and conservative of the Polynesian cultures (e.g., Hanson, 1973; Holmes, 1980b; Stanner, 1953). As we see later, however, the missionaries did introduce significant changes into Samoan culture by altering the way the young are socialized and recruited into adulthood. They learned very early that the way to change a culture is to alter the way the young "come of age."

The previous description of precontact Samoa presents a picture of well-structured socialization processes governing the coming-of-age process. In particular, the *'aumaga* and *aualuma* should have ensured that behavior was quite well regulated among the young. And these processes seem to have worked well, because the culture has survived for centuries more or less in the form encountered by the missionaries. Although undoubtedly some young people found these processes stressful and some probably engaged in what we call delinquency, if life had been as stressful and delinquency-prone as Freeman argued, there would be far more evidence of such problems. For example, the early missionary

ANALYSIS

accounts make no mention of either excessive stress or delinquency. Had there been problems in these respects, the missionaries surely would have given them as much attention as they did other "evils." If anything, the earliest records emphasized the opposite in terms of the quality of life available (cf. Meleisea, 1987a, p. 34).

In addition to these structures, precontact Samoan culture had other characteristics and mechanisms that would have minimized behaviors that might have threatened community integrity and cultural stability. An examination of these confirms the view of a well-regulated society—one that would remain intact over dozens of generations.

As we see, precontact Samoan communities employed a number of mechanisms that were apparently effective in minimizing the incidence of social deviance, and remedying much of the damage that might have been caused by this deviance. Again, this is not to say that precontact Samoa was a paradise. Most notably in contrast to this image were the wars that apparently occurred with some frequency (Freeman, 1983). But, again, our concern is with the institutions that would have consistently affected how the young would come of age, and to these we look *within* communities, not between them. There is no indication that these institutions were affected by any constant war-readiness.

First, villages averaged about 350 inhabitants. Thus, precontact Samoan culture did not have to deal with the types of deviance associated with anonymity or excessive population density. Moreover, as Gilson (1970) described, any threat to village integrity was dealt with immediately:

> when a village was irreconcilably divided on any important issue, local or otherwise, the weaker party might be driven out in order to restore village unity. Individuals who committed serious offenses were treated no less harshly. When travellers or visitors stirred up trouble or acted disrespectfully toward the village, even unintentionally, they were liable to immediate attack and expulsion by the *'aumaga*. The Samoans were so sensitive about village honour and the enforcement of local customs and regulations that their reaction against transgressors was inclined to be hasty and violent. (p. 48)

Second, a well-defined authority structure (the *fono*) existed that was replicated in all villages. Part of its mandate was to coordinate relationships among villages. The *matai* was the undisputed head of the *'aiga*, which could include more than 60 persons. Membership in the *'aumaga* required the observance of the principle of the *tautua*; namely, service to the *matai*. Through this service, young men gained respect and status in the eyes of the adult community, and could perhaps eventually earn a *matai* title. Associated with this service was the requirement that the authority of the *matai* not be challenged. If dissatisfaction over some thing was felt, it was a "privilege" to raise it with the *matai*, not a "right" (MacPherson & MacPherson, 1985).[5] The decision of the *matai* on the matter was final, however, and any questioning of the decision was considered childish (*fia pepe*) or immature (*ma fau*) (MacPherson & MacPherson, 1985). Physical punishment of young people occurred when other avenues of control failed, but was apparently required on an infrequent basis (Holmes, 1987).

Third, if a young person found his or her *matai's* decisions unacceptable, or day-to-day living a problem, he or she could sometimes change households within the village or go to another village. "Adoptions" were common for younger individuals, but anyone appearing in an *'aiga's fale* (house) who was in need of shelter was taken in (cf. Schoeffel & Meleisea, 1983).

Finally, to remedy damage caused by more serious forms of deviance committed between *'aiga* (and villages), an apparently effective mechanism referred to as the *ifoga* was employed. The *ifoga* seems to have provided a means of shaming the offender as well as producing a reparation that was not too disruptive of community life. According to this custom, the offender and his or her *matai* would go to the doorstep of the offended *aiga* and ask for forgiveness. They would do so early in the morning, cover themselves in "fine mats" and stay there all day, or for several days if necessary. This would be done until either forgiveness was forthcoming and they were invited into the household, or physical revenge was taken on them (Filoiali'i & Knowles, 1982). This custom survives today, although a Western-style legal system now officially attends to matters of justice.

Given these cultural characteristics and mechanisms, it is unlikely that serious delinquency, or other forms of seriously disruptive deviance, existed to any great extent. Readers are reminded of the examination of Freeman's statistical and anecdotal evidence in chapter 2. There it was concluded that even if "high" cross-culturally, the various forms of social deviance were likely sporadic occurrences in a given village. Of course, there are no written records from precontact culture to quantify these behaviors. In light of this discussion, however, it is *implausible* that village life was highly problematic or stressful, particularly on the small, isolated island of Ta'u.

SEXUAL PRACTICES IN PRECONTACT SAMOA

The issue of Samoan sexual habits has been associated with Mead's book since it was first published. Soon after it was released, some critics called it scurrilous for the attention given to the sexual behavior of young women

---

[5]This is a recent source describing a surviving convention, but the convention appears to be so fundamental to *fa'a Samoa* that it undoubtedly predated Western contact.

(Cassidy, 1982). A. Wendt (1983) believes that the "instant worldwide popularity of Mead's Samoa was due largely to her delectable portrayal of free love without guilt under the huge swaying palms" (p. 10).

As we saw in chapter 1, this is the issue upon which many Samoans have focused. Consequently, many view Mead's work with contempt because of its incongruence with contemporary Christian Samoa. Unfortunately, this issue has overshadowed the many other important matters raised by Mead's work. Thus, the current mythology encourages the belief that the focus of her book was about adolescent sexual habits rather than about coming of age.

Despite what some contributors of the Mead–Freeman controversy have claimed, the issue of sexual behavior is not crucial to Mead's coming-of-age thesis. Furthermore, she never asserted in *Coming of Age* that casual sexual relations were a necessary component to an ease of coming of age. In fact, it is plainly there for anyone to read that of "the twenty-five girls past puberty, [only] eleven had heterosexual experience" (Mead, 1928, p. 151). Because she argued that the other 14 also had an ease of passage, she obviously did not see sexual experimentation as a necessary component of this ease of passage. As we see in chapter 6, she found sexual experimentation to be an *option* for some, not an expectation or a requirement. On the other hand, as we see later, it is likely that young Samoan females had more sexual freedom before Mead's time—before they lost rights in the *aualuma*, and before many were sequestered into the pastor's house where their chastity was required.

Therefore, although it is actually tangential to Mead's coming-of-age thesis, I am taking the time to deal with the issue of sexuality in this book because of the emotionally charged mythology surrounding it. Freeman brought worldwide attention to this issue, probably to the horror of some conservative Samoans, by arguing that Mead was wrong in her accounts of premarital sex among her informants because she took seriously a teasing (or "duping") by her informants on sexual matters (Freeman, 1983, pp. 289–290, 1989a). By examining the available writings about the matter, the issue can perhaps be put to rest. At the very least, the concern over Mead's credibility can be addressed, given that her book was made possible by her rapport with these young women. Accordingly, it is worthwhile to consider whether her reports of such behaviors are plausible.

We begin this analysis with evidence provided by the first missionary to make written records of precontact Samoa—John Williams (cited in Moyle, 1984).

Williams found a Samoa in which polygamy was widespread. In his 1830 and 1832 journals he pondered over what to do about polygamy because it was "inconsistent with Christianity" (p. 142). However, he did not want to immediately impose the Christian practice of monogamy because it was mainly the chiefs who practiced polygamy—"some have two others have as many as six" wives (p. 142)—and if he threatened the chief's interests they might turn against him and Christianity. His dilemma was expressed in the following way:

> it is . . . a subject fraught with perplexity. In many instances a man has two wives with whom he has lived for many years & has a family by each feels an attachment to both, & of course is at a loss which to put away. If he is allowed to keep his two others will make it a pretext for keeping half a dozen so that prudence & consideration must be exercised in the management of this affair. I have thought that in some instances perhaps it might be well to allow them to retain their wives & yet admit them into Christian communion, *but the wild young Chiefs who have so many wild young girls to none of which he is lawfully married according to the customs of the country* of course the putting of them away would be attended with no difficulty for that is frequently done. For when a young Chief wishes to marry a *maitai* or lady she will not hear of his proposition until he puts away all his other wives. This he does without hesitation. (pp. 142–143, italics added)

Actually, according to Meleisea (1987a), the missionaries' social engineering was more complicated than Williams thought because the

> changing of one custom often led to a series of changes. For example, . . . [c]hiefs wishing to become Christians were asked to choose one wife for the rest of their lives . . . Since the main reason for chiefs contracting so many marriages was the Samoan political system. . . . changing that custom also greatly changed the political system of Samoa. It was another eight years after 1830, before Samoans really accepted the new teachings on marriage, and when they finally did, the conferring of *taupou* titles became less important. This is because the institution of the *taupou* was closely associated with chiefly marriage, which linked families and villages all over Samoa through the multiple marriages of chiefs to high ranking ladies. (p. 68)

With respect to general sexual behavior, Williams recorded the following impression of the Samoans he encountered in the early 1830s: "I should certainly on the whole consider them a people of lascivious habits far more like the Tahitians than the Tagnatabuans to whose female population the palm of modesty must be conceded by all the ladies of the Eastern groups even to the Sandwich Islands" (p. 233).

On the matter of premarital sex, Williams recorded a practice that institutionalized a free sexuality for unmarried females. He described this practice as follows:

> It is also a common thing for [unmarried] young women to be publicly deprived of their virginity by a young respectable Chief in the same way

as at the marriage ceremony [a public, manual rupturing of the hymen]. This is considered an honour & no person objects to marry a young woman who has been thus treated. The Chief who ruptures the Hymen will frequently give the young woman a great name which will gain her respectability but *I suspect the reason why this singular system prevails is the young females are tired of submitting to the restraints their virginity imposed on them & by being thus honourably deprived of their virginity they have full liberty to gratify their wishes & also escape the disgrace of being looked upon as common prostitutes.*

The Teachers inform me that six young females had been thus treated within a short space previously to my arrival by the son of Malietoa. These occasions are always attended with dancing & merriments of every description. (pp. 256–257, italics added)

This passage suggests a *tolerance of premarital sexuality even greater than that reported by Mead*. This would make sense given that Mead's observations were made after some 80 years of Christian influence. But, perhaps the greatest corroborating evidence for Mead's account comes from Williams' statement that "[t]he young females I have heard have a great objection to marriage[,] a roving commission being more congenial to their feelings" (p. 233). As we see later, Mead (1928, p. 195) also indicated that some of her informants reported wanting to delay marriage as long as possible, and instead engage in casual sexual encounters.

Williams continued by noting that although he was told that adultery was rare and was punishable with death, he was skeptical. The reader will note here that the ideal of chastity apparently existed then, but was also likely violated regularly.[6] Williams stated the reason for his skepticism that the cultural ideal actually translated into behavior:

I am not a convert to this opinion for I know too much of the natives of the South Seas generally to allow me to suppose that chastity could be maintained under the circumstances in which the Samoa females live. It is a very common thing for one Chief to have from two to six or more wives. The rule established is that each female is to lie with husband three successive nights. Now allowing that the husband has six wives it will be fifteen nights again before she is admitted to his bed. During these fifteen nights she sleeps where she likes or generally speaking sleeps in an adjacent house where all the persons of the establishment men & women sleep promiscuously together under one cloth so that at least there are fifteen chances against the alleged chastity of some. (pp. 233–234)

---

[6] He suspected that when punishment was meted out it was likely only for those who had violated some other norm or who threatened the interests of certain chiefs. Western culture had a similar practice with the burning of witches in England and the United States (Karlsen, 1987).

A number of other entries made by Williams gives us some idea of the social atmosphere that might have prevailed in 1830s Samoa with respect to sexual mores. For example, in the following account, we glimpse something of the precontact idea of decorum in formal dining among chiefs and their honored guests:

> This afternoon I had the honour of being surrounded at tea time by his Majesty's five wives three of which are about five & forty or fifty years of age the other two are girls about seventeen or eighteen. On being invited to partake they drew nearer & formed a circle dropping their little round abouts off their loins on to the ground. They curled the small end of it into the lap just barely concealing the only part that they even make a pretension to conceal. Having asked a blessing they eat heartily and cheerfully of what was placed before them. This is the first time I have had the honour of eating with *five naked* queens. The natives appear to take the greatest pride in exposing their persons. (pp. 143–144)

Obviously, Samoan conceptions of modesty were quite different from the conception Williams took with him from his native land. We see this in more detail in the following passage:

> The females however are by no means as careful about concealing their persons as the men. At the Samoa Islands the care is on the part of the men. I have seen a man demand from a female a lavalava or round about to cover his person while she is obliged to go naked. Clothing of every kind appears a burden to them. If a person should jeer a young woman as she might be passing by remarking freely on her person saying she was diseased or ill formed she would instantly throw off her cloth & expose herself in every possible direction & pass on. A respectable young man who had been residing among them some time informed me that when he first went on shore among them the females in great numbers gathered round him & some took off their mats before him exposing their persons as much as possible to his view. Perceiving him bashful the whole of the women old & young did the same & began dancing in that state before him desiring him not too bashful or angry it was Faa Samoa or Samoa fashion. (p. 232)

Thus, from Williams' journals it appears that the standards of modesty, decorum, and sexual conduct obtained in contemporary Christian Samoa would not be found in precontact Samoa. In fact, as we see in chapter 5, the missionaries worked for decades to "Victorianize" Samoans and it took generations to totally eradicate some practices. Accordingly, one of their earliest objectives was to convince Samoans to wear clothing: "Some few Samoans who have embraced Christianity have taken to wear cloth entirely. On Sabbath days also the Teachers have succeeded in inducing the whole congregation men & women to attend properly clothed & decently covered" (p. 231).

Turner (1861/1986) also provided us with records of Samoa before Christianity was more fully adopted (the 1840s and 1850s). With respect to premarital sex, he wrote the following:

> Chastity is ostensibly cultivated by both sexes; but it is more a name than a reality. From their childhood their ears are familiar with the most obscene conversation; and as a family, to some extent, herd together, immorality is the natural and prevalent consequence. There are exceptions, especially among the daughters of persons of rank; but they are the exceptions, not the rule. Many native teachers and other consistent characters, seeing the evil, have now separate sleeping apartments in their dwellings; and their better regulated families are becoming models to their countrymen of an improved and improving community. (pp. 90–91)

Note in the latter portion of this account that the missionaries by this time were beginning to have an impact and that some Samoans were ostensibly altering their behavior, at least for the sake of the missionaries. With respect to extramarital sex, Turner wrote the following:

> Adultery, too, is sadly prevalent, although often severely punished by private revenge. If the injured husband seeks revenge in the blood of the seducers, no one thinks he has done wrong. But the worst feature of the law of private revenge is, that the brother, or any near relation of the culprit, is as liable to be killed as he himself is. Fines are now being substituted; but occasionally, revolting murders are committed on account of the crime. (p. 91)

Turner recorded the emotion of jealously and the custom of revenge for adultery, matters that Freeman raised. Undoubtedly, these were components of precontact Samoan culture, but it does not appear that they were simple matters. Indeed, the nature of jealousy and the definition of adultery both appear to be situational in nature and are dependent on the nature of the agreement between the couple. Thus, in explicitly monogamous relationships, these prescriptions probably prevailed for most couples, and the missionaries undoubtedly cultivated the emotion of jealousy to increase the chances that couples would stay in "Christian unions." For polygamous unions, however, other standards likely applied. For instance, if we return to Williams' records, we find an account of the following practices:

> The husband appears passionately fond of his wife for a short time. Sometimes it will continue for a year or two sometimes until she becomes preg-

nant. Generally at the end of a year or eighteen months the wife will not only give her consent but expect her husband to take another female as wife as well as herself, & it frequently occurs that the wife herself will go to her own tribe & seek from her friends & relations the female who is to share with her affections of her husbands. *This perhaps may be one reason why generally speaking copartnership in husbands by Samoan females is attended with so little jealousy & quarrelling.* The universal rule established among them of giving to each wife in rotation her three days & three nights supremacy may also tend to promote union & good feeling among the wives toward each other. (p. 258, italics added)

Another account from this period of Samoan sexual behavior comes from Stair (1897/1983), who lived in Samoa from 1838 to 1845. Stair's account verifies Williams' record:

Polygamy at one time was largely prevalent, many chiefs having seven or eight or even more wives at a time. There were generally one or more principal wives, who kept the others in subjection, often exerting their authority with a very high hand. Formerly, a discarded wife of a chief, or one who had voluntarily left her husband, was prohibited from marrying another man, unless the latter were powerful enough to set this prohibition in defiance. As illustrating a custom common at one time in Samoan society, I may mention that many women have assured me that when it is seen that the husband was resolved upon adding another wife to his harem, the principal wife often selected her own sister or sisters, and endeavoured to get them added to the family roll of wives, so that she might have some control over them. This plan was frequently adopted to avoid strangers being brought into the family. (p. 175)

Finally, Williamson (1939/1975) carried out an extensive review of all of the early accounts of Polynesian cultures, including Samoan culture. Based on these varied sources he found it reasonable to draw the following conclusions: With respect to virginity, he stated that like other Polynesian cultures "virginity is a social asset rather than a moral virtue, and is rigorously preserved only in the case of the taupou [ceremonial village virgin]" (p. 324). With respect to premarital sex in general, he said that in Samoa:

According to Turner and Brown, chastity in either sex, prior to marriage, though ostensibly cultivated, was more of a name than a reality, and the lack of it was not regarded as a very serious offence; but, according to Turner, exceptions to this general rule were found, especially among the daughters of persons of rank. D'Urville says that girls were entirely free to dispose of their persons till married, and La Perouse tells us that girls were, before marriage, mistresses of their own favors, and their complaisance did not dishonour them. (p. 156)

Williamson also found evidence of the complexity of the Samoan mores discussed by Schoeffel and Meleisea (see chapter 3):

> A ceremony on marriage for testing the virginity of the bride was . . . performed not only in the case of daughters of chiefs, but in that of the commoner people also; and even among the latter, a girl who did not emerge successfully from the test was regarded as disgraced, and might be beaten; though the disgrace to her seems to have been nothing compared with that of a chief's daughter, who might be killed. This suggests that immorality, even among the humbler classes was not as general as some writers make it out to have been; *but it cannot be doubted that girls of these classes were much more lax than were the daughters of chiefs.* (pp. 156–157, italics added)

Although it is possible that some women were punished or killed after they failed the test of manual defloration, it appears that provisions could be made so that even nonvirgins could pass the test. As Holmes (1957b) said, "informants state that many a girl has been saved embarrassment by the substitution of a chicken bladder full of blood for that normally produced by the broken hymen" (p. 413). Indeed, much of the anger at someone failing the test would be over the foolishness of submitting to the test unprepared when one was not a virgin, because this would cause shame to the *aiga* and invalidate any economic arrangements surrounding the marriage. On this matter, Mead (1928) wrote that

> [a]lthough the virginity-testing ceremony was theoretically observed at weddings of people of all ranks, it was simply ignored if the boy knew that it was an idle form, and "a wise girl who is not a virgin will tell the talking chief of her husband, so that she be not shamed before all the people." (p. 98)

In reference to polygyny, Williamson noted that the practice, common among chiefs, was reported by most observers:

> There were generally . . . one or more principal wives, who kept the others in subjection, often exerting their authority with a very high hand; and when the husband proposed to add another wife to his establishment, his principal wife often secured the selection of her own sister, thus structuring her control and avoiding the introduction of strangers into the family. When a woman married (presumably only one of the upper or wealthier classes), it was usual for her to take with her to her husband one, two, or three concubines. . . . The concubines, though only regarded as secondary wives, were not usually despised, and any of them might eventually be taken as a wife without further ceremony. Kramer explains it by saying that it was not usual for a woman to return to her husband for six months after the

birth of her child; so she brought with her at marriage some of her women relations, to prevent her husband from going to other women during the interval of separation. (pp. 178–179)

From these many accounts, there can be little doubt that sexual behavior in Samoa before it was Christianized was more casual for virtually *everyone, including young females* (see also Howard & Kirkpatrick, 1989, pp. 78–84). The denial of this by Freeman and some contemporary Samoans can be understood in terms of the concerted efforts of missionaries and the local pastors to first create, and then maintain, a hegemony of Victorian sexual values and practices. These efforts have been so successful that the very thought of precontact "promiscuity" appears to be anxiety provoking for many Samoans, so various defenses are employed to deny it.

Freeman's "mission" can be understood as part of this collective denial. In fact, his own reconstruction of precontact behavior is highly selective (see Freeman, 1983, chapter 16). For example, he cited a portion of a sentence from Williamson's account (that "virginity is a social asset rather than a moral virtue"; 1983, p. 230), but he conveniently omitted the remainder of the sentence; namely, that virginity "is rigorously preserved only in the case of *taupou*." Freeman made no mention of the additional documentation provided above from Williamson, nor did he mention the passages from Williams, Turner, or Stair. Furthermore, Freeman's own conduct in this matter stands as starkly hypocritical when we find that he has admonished Boas for not verifying Mead's account by checking these sources. In his words:

> What can be said with certainty is that if Boas, as the instigator and supervisor of Mead's Samoan researches, had taken the elementary precaution of consulting the readily available ethnographic literature on Samoa, as, for example, the writings of *Williams, Turner*, Pritchard, Stuebel, and Kramer, he would have very quickly found accounts of the sexual and other behavior of the Samoans that are markedly at variance with Mead's picture of life in Manu'a. . . . (Freeman, 1983, p. 291, italics added)

CHAPTER FIVE

A Social History of Adolescence in Samoa: Changes in Samoan Culture

MISSIONARIES AND THEIR IMPACT

Missionaries like George Turner and John Williams played a key role in initiating the changes wrought by the West on Samoan culture. Conveniently, they kept records of their perceptions regarding what they found, and of their intentions regarding what they set out to change, so we can now trace their influences on contemporary Samoa.

When reviewing Turner's (1861/1986) record, one is struck by his disdain for the culture he found in Samoa. For example, he (and many other missionaries) viewed many Samoan customs as "evil," so he set out to eradicate that evil and to teach the "heathen natives" (p. 4) to live by what he believed to be the morally superior principles of Euro-Christian culture. Moreover, in what some would today consider delusional, Turner believed that "Satan . . . [held] . . . dominion in Samoa [and he was not going to give it up] without a struggle" (p. 7). Turner described several "schemes" (p. 9) that he believed Satan had devised to prevent Samoans from embracing Christianity. Here, we examine Turner's struggles with Satan and the negative impact of the outcome of these struggles on young Samoans. First, it is enlightening to place the whole of the missionary effort in a sociological context.

Gilson (1970) provided a thorough review of the changes that took place in Samoa during the 19th century. Included in this review is the following summary of the agenda of the London Missionary Society

(LMS),[1] the strongest Christian influence in Samoa and the sect that prevailed on Ta'u when Mead was there. According to Gilson, the ultimate goal the LMS set for itself in Samoa was "to ban the activities and relationships, social and personal, that by mission standards were immoral or tainted with 'heathenish' associations, and to prescribe the ethics and conventions of puritanism" (p. 96). In reaching this goal, many Samoan customs had to be eradicated. Given priority in their agenda was the "Christianization" of "sex and family relations," including the following:

> the abolition of polygamy and, in most cases, divorce, the celebration of monogamous marriages in church; the prohibition of certain customary marriage rights, including the exchange of goods and the public test of virginity; the prevention of political marriages and of marriages between Christian and non-Christian; the prohibition of adultery, fornication and prostitution; the prohibition of obscenity in word and action; the imposition of new standards of dress, including "full coverage" for women and, when at worship, shirts or coats for men, but not shoes for either; the adoption of hair styles "appropriate" to the individual's sex, meaning long for women and short for men, the reverse of traditional styles; [and] the internal partitioning of houses and more liberal use of the external blinds. (p. 96)

Also given priority was the inculcation of "Christian" values and habits. The most extensive of these included the insistence that "the Sabbath should be strictly kept and that *habits of industry* should be encouraged. In the latter aim was enshrined the *dignity of work*, and with it the assurance that the people would be able to clothe and house themselves properly and contribute to the *support of their mission and church*" (p. 97, italics added). We see this agenda in the following passage taken from Turner (1861/1986). Here he described his experiences in attempting to instill the values of individualism among Samoans:

> The system of a common interest in each other's property . . . is clung to by the Samoans with great tenacity. . . . This common property system is a sad hinderance to the industrious, and eats like a canker-worm at the roots of individual or national progress. No matter how hard a young man may

---

[1]The London Missionary Society (LMS) was a nondenominational Protestant sect whose sole reason for existence was "to spread the knowledge of Christ among heathen and other unenlightened nations" (Moyle, 1984, p. 2). Initially, most of the financial support for the LMS "came from the Congregational Church with whom, in later years, the [LMS] became exclusively associated. The Congregational Church was different from other non-conformist churches because it preferred decisions concerning church government to be made by the congregation rather [than] by bishops or ministers" (Meleisea, 1987a, p. 55). The LMS "created the first independent church in the South Pacific with a fully indigenous ministry and administration. Today it is called the Christian Congregational Church of Samoa" (Meleisea, 1987a, p. 60).

be disposed to work, he cannot keep his earnings: all soon passes out of his hands into the common circulating currency. (p. 170)

It is not clear what the "national progress" was that Turner was thinking of for these island-bound people who had done quite well for centuries before his arrival. Surprisingly, however, Turner did display an inkling of comprehension of how Samoans recognized the benefits of their communal system:

> The sick the aged, the blind, the lame, and even the vagrant, has always a house and home, and food and raiment, as far as he considers he needs it. A stranger may, at first sight, think a Samoan one of the poorest of the poor, and yet he may live ten years with that Samoan and not be able to make him understand what *poverty* really is, in the European sense of the word. "How is it?" he will always say. "*No food!* Has he no friends? *No house to live in!* Where *did* he grow? Are there no houses belonging to his friends? Have the people there no love for each other?" (pp. 170–171)

Thus, included in the missionary agenda was the goal of diminishing the Samoan concern for collective welfare, and replacing it with Western individualism. As someone who was raised to take the teachings of Christ seriously, I find this part of the agenda to be a curious hypocrisy for those claiming to be emissaries of Christ. But, then, the (largely Protestant) missionaries functioned as the "shocktroops" for the onslaught of Western economic practices based on the profit motive, such as wage labor, "middle class individualism" (Meleisea, 1987b, p. 18), and careerism. Samoans were told that initiative and individual striving were what pleased God, not looking after each other's welfare; they were also told that their primary concern for the collective was misguided, although it was acceptable as a subsidiary social concern. Meleisea (1987a) framed the missionary mind-set in the following manner:

> The first missionaries to Samoa were of the lower middle classes of England, the backbone of the Evangelical movement there. They believed that hard work was the duty of a Christian. They thought Pacific Islanders lazy because the hours that people worked were so different to those of the English. For example, Samoans worked very hard at times of planting and harvesting, but there were also times when there was plenty of leisure. The missionaries wanted to make Samoans more industrious. One way of doing this was by introducing new goods to the people which necessitated their working for wages, or selling products, in order to obtain them. The idea which the missionaries had of the way Samoans should live, was based upon the ideals of the middle classes of England; where households consisted of the married couple and their children. The man was the breadwinner, and produced the food, working regular hours outside the home.

The woman was the home-maker, and cooked and sewed clothes in the house for the family. (p. 67)

On point after point, the missionaries told Samoans that their customs were sinful and that they should feel guilty when they behaved in ways that offended God. These were not easy things to impress upon people from a culture that did not recognize one God, let alone one that was so punishing and vengeful when offended (cf., Holmes, 1987, pp. 59–60). Observers agree that most Samoans remained ambivalent to the Protestant work ethic and material accumulation, something that continually frustrated Western entrepreneurs (Oliver, 1961). But, guilt-driven motives were not easy notions to instill in a shame-based culture (cf. Gerber, 1975).[2] In fact, this latter point may help explain why many of the changes instituted by missionaries seem to be more a matter of word than deed. The strong evidence of "dualism" discussed in chapter 3 supports this notion; namely, that in certain matters some Samoans may go through the motions of conformity to avoid being shamed, but not feel guilty for "discrete" violations of Christian conventions, such as breaking the convention of sexual abstinence.[3]

Completing the agenda for the Christianization of Samoa was to be no small task, largely because of the resistance of Samoans, so the missionaries gave themselves generations to accomplish it. The missionary agenda ultimately succeeded in converting Samoans to a *form* of

---

[2]The concepts of *sin* and *guilt* appear not to have had wide currency in the traditional culture. Indeed, it appears that the missionaries promoted these concepts, with some difficulty, as described later. Instead of guilt, the primary emotion motivating approach-avoidance behavior seems to have been shame. Thus, if one violated a norm, one brought shame onto one's self, but also onto one's *aiga* and one's *matai*. Given the fact that privacy was not valued and that little norm violation went unnoticed, this would constitute a stronger social control mechanism than guilt, as employed by Western culture, because guilt is much more private and can be controlled psychologically (although with "neurotic consequences"). Moreover, the *fono* would levy fines against the offender's *aiga* and *matai*, not just the offender. Therefore, would-be offenders would know that they would have to account for their behavior to their extended family as well as the village.

[3]As a point of interest, Somerset Maugham (1985) provided a caricature of this attitude in his famous short story, "Rain," set in Samoa:

> "When we went there they had no sense of sin at all," [Reverend Davidson] said. "They broke the commandments one after the other and never knew they were doing wrong. And I think that was the most difficult part of my work, to instill into the natives the sense of sin." (p. 256) "You see, they were so naturally depraved that they couldn't be bought out of their wickedness. We had to make sins out of what they thought were natural actions. We had to make it a sin, not only to commit adultery and to lie and thieve, but to expose their bodies, and to dance and not to come to church. I made it a sin for a girl to show her bosom and a sin for a man not to wear trousers." (p. 258)

Christianity—the hundreds of churches that are the center of communities throughout the islands are a testimony to this success. Yet, in spite of 150 years of concerted effort, many Samoan traditions have survived in some form and many Samoans are still proud of this *"fa'a Samoa"* (the Samoan way of life). See Holmes (1980a, 1987) and Meleisea (1987a, 1987b) for analyses of how and why Samoans embraced Christianity.

The complexity of Christian-Samoan culture can be attributed in part to the fact that the conversion of Samoa was the culmination of numerous *rational* choices. These choices were made in the *fono* of virtually every village in Samoa. Therefore, the view that Samoans were passive throughout the conversion process is a simplistic one. The decisions to embrace Christianity was often based on the promise, however implicit, that material advantages would follow. As Holmes (1987) argued, in "exchange for all [the] prohibitions the Samoans . . . probably expected . . . that ultimately they would also acquire some of the white man's magic for getting material wealth" (p. 61). Consequently, contemporary Samoan life is more complex than it may appear based on a superficial analysis, and some practices continue to exist along side of Western *religious* conventions (as with tattooing) and Western *legal* conventions (as with the *ifoga*) (cf. Hanson, 1973; Holmes, 1980b; Meleisea, 1987a; Oliver, 1961; Stanner, 1953).

Without this complexity in Samoan culture, I doubt the Mead–Freeman controversy would have emerged at all. Although this complexity is confusing to outsiders, it also seems to have become a source of confusion for many Samoans, particularly those now attempting to come of age. For those interested in pursuing this matter, I highly recommend A. Wendt's (1973, 1977, 1979) novels and short stories, in which he portrays the complexities of 20th century Samoan culture and the resulting confusions experienced by some Samoans.

MISSIONARY IMPACT ON COMING OF AGE

The missionaries' desire to recreate Samoan culture in the image of Victorian-European culture represents a distinct lack of sensitivity to the validity, and an appreciation of the beauty, of that culture. Holmes (1987) noted in this regard that "the early English missionaries seemed to be unable to separate their religious beliefs from their culture and values" (p. 60). As a part of their "mission" to Christianize Samoa (and the world) and as a consequence of their insensitivity to the institutional balances in the culture they were transforming, many of the changes instituted by the missionaries have proven to have been harmful to Samoans, particularly young Samoans. From the point of view of the interests of the young people who are inheriting the transformed culture, the influences of the missionaries are harmful to the extent that they destroyed some of the best traditions of Samoan culture and allowed them to be replaced with some of the worst practices of European culture.

These criticisms of missionary activities are not intended as a claim that missionaries are solely responsible for all that has happened to Samoan culture since the 1830s. Nor am I denying that the missionaries had what many would see as a positive impact, such as health care and literacy. Nevertheless, to complete their "mission," they felt they had to eradicate "heathenism," and to do so it was necessary to transform the culture. It appears that their strategy of culture transformation was to target how Samoans would come of age. Had they preserved the existing institutions governing coming of age, the society would have remained largely intact. As indicated earlier, the surest way to change a culture is to change what the young are taught and how they are socialized—thereby changing what they *value*. Accordingly, as we see in this section, the missionaries set out to destroy most of the social institutions that guided young Samoans through childhood to adulthood. In doing so, they also diminished the status of young people in Samoan society—culturally disenfranchised them—because the institutions they destroyed had guaranteed them basic rights as persons. With diminished status and rights, many of those coming of age have had a lower sense of purpose, more difficulty developing a sense of identity, and less affinity with their culture. I explore the ramifications of these problems in chapter 7. For now I focus on the specific changes initiated by the missionaries.

One of the missionaries' first objectives was to eradicate the practice of tattooing. Turner (1861/1986) provided the following justification for doing so—note the disregard of the fact that tattooing had marked a rite of passage for Samoan youth for several centuries:

> The waste of time, revelling, and immorality connected with the custom have led us to discountenance it; and it is, to a considerable extent, given up. But the . . . youth still thinks it is manly and respectable to be tattooed; parental pride says the same thing; and so the custom still obtains. It is not likely, however, to stand long before advancing civilization. European clothing, and a sense of propriety they are acquiring daily, lead them to cover the tattooed part of the body entirely; and, when its display is considered a shame rather than a boast, it will probably be given up, as painful, expensive, and useless; and then, too, instead of tattooing, age, experience, common-sense, and *education* will determine whether or not the young man is entitled to the respect and privileges of mature years. (pp. 89–91, italics added)

Given his disdain for the practice among men, Turner must have been horrified with the tattooing of women. Undoubtedly, the tattooing of women would have been considered even more "evil" by the early missionaries because of its immodesty (see Fig. 4.1), and their efforts were likely more concerted in the case of women. Although the practice seems to have been consequently eradicated among women, it was to be impossible to eliminate among men. As late as the 1920s, many missionaries were apparently still consumed with the notion of abolishing it. This was recorded by the New Zealand administrator Rowe (1930), who noted that "the missionaries—with the exception of the Catholics—hated it, and still hate it as a relic of 'heathenism.' It matters nothing apparently to them that, while the custom stands, it militates against immature mating; and that *it is the one test in these islands, where life is so easy, that the youth has to go through*" (p. 85, italics added).

Note from the emphasized portion of Rowe's passage his intuition that the attack on this practice was negatively affecting youth. Rowe continued by noting that the best tattooers were on Savai'i, where missionary influence was weaker, and that "the population of Manu'a was likely to be reduced by this cause [getting tattooed], as those who indulged in the fashion were forbidden by law to return" (p. 73). It is of interest to note that this coincides with a minor detail from Mead's (1928) study, which was conducted during the same decade. She reported that "[t]attooing has been taboo on Manu'a for two generations, so only part of the population have made the necessary journey to another island in search of a tattooer" (p. 267).

The situation in contemporary Samoa is proving Turner wrong regarding the entitlement of respect and privilege that accompanies "advancing civilization." For example, for many young people in Western Samoa, education does not now "determine whether or not the young man is entitled to the respect and privileges of mature years" in their culture (Turner, 1986/1861, pp. 89–91). Rather, education seems to drive a wedge between many young people and their ancestral culture. As for tattooing, this practice is enjoying a resurgence in popularity, probably as a direct response to the deteriorating circumstances confronting young males. Indeed, recently, "more and more [young men] are deciding to undergo this painful procedure to display their Samoan identity" (McGovern, 1988, p. 28; see also Holmes, 1987, p. 76).

O'Meara (1990) suggested that tattooing is popular today partly because:

> the tattoo is quintessentially Samoan, and wearing one is a matter of cultural pride.... The tattoo ceremony also marks the beginning of adulthood. In addition, *it brings a great deal of attention and praise to young*

men at a time in their lives when they normally get very little of either. (pp. 75–76, italics added)

He also documented some of this recent resurgence on the island of Savai'i. In addition, he reported that tattooing was still common there in the 1920s and a revival in the practice began about the time of political independence for Western Samoa (in 1962). The tattooing of women there is now rare, but most men have it done in "their late teens or early twenties" (p. 72). Other than doing it a few years later than was the tradition (age 16), much of the traditional procedure is followed. Those who are ready, organize themselves into a small, mutually supportive group. Because it is a long and painful process, sharing the experience not only provides emotional support, it also undoubtedly creates a strong bond among those undergoing it. From this bond, a common identity with each other can be derived, based on shared roots within the ancestral culture.

Although the struggle over tattooing interfered with the desire of young Samoans to display their adult status, a more serious struggle took place that removed their opportunity to come of age in a relatively continuous and stress-free fashion (cf. Mead's coming-of-age thesis). This was the assault on the *aualuma* and the *'aumaga*.

The *aualuma* was the first to go. The missionaries disbanded this institution as soon as was practicable. Subsequently, "unmarried girls, instead of living with the *aualuma*, had to live with the pastor and his wife where they learned to cook in the papalagi [European] way, using indoor stove and pots and pans, as well as learning sewing and other papalagi women's work" (Meleisea, 1987a, p. 68). Holmes (1957b) provided the following account of the demise of this social institution:

> Formerly, a girl of fourteen or fifteen would be ready to enter the group of unmarried girls (*aualuma*) who slept together and formed the entourage of the village ceremonial virgin (*taupou*). Traditionally this group entertained ceremonial visiting parties (*malaga*), carried out village work projects and served as aids to the *taupou*, who served the village in the capacity of ceremonial hostess and dance leader. The *taupou* system has greatly degenerated since the turn of the century. (p. 409)

Holmes (1987) also observed that the *aualuma* "is now but a part of a greater village organization known as the Women's Committee. This group is made up of the wives of untitled men and the wives of Chiefs and Talking Chiefs" (p. 43). Keesing (1934) provided the following description regarding how the Women's Committee came into being and usurped the mandate of the *aualuma*:

Village Women's Committees were first formed under medical auspices in the 1920s to improve infant care, health, and sanitation. During the subsequent antigovernment *Mau* ("Opinion") movement, men's assemblies above the village level were officially forbidden, and women for a time assumed the overt exercise of political functions, with assemblies all the way up to *Mau* Women's "Councils of all (Western) Samoa." Subsequently, women's committees withdrew once more into activity mainly concerned with hospitality, health, child care, village cleanliness, and hygiene, though with greatly enhanced authority and dignity. *The former aualuma institution has been incorporated as a kind of "junior league" at the lower rungs of the powerful feminine elite ladder.* Women's Committees in Western Samoa have local uniforms, usually brilliantly coloured, and very recently the idea of building a special house in each village as a center for their activities has been spreading throughout Western Samoa. (p. 56, italics added)

Thus, the principal social structure that helped to guide young Samoan women to adulthood has been drastically transformed and now is a mere appendage of an adult-led and adult-oriented organization. In contrast, in precontact Samoa, the unmarried women of a village would live together, separate from adults. When *'aumaga* from neighboring villages visited (*malaga*), "[a]fter feasting and dancing, *the two groups slept together* in the house set aside as the residence of this female group (*faleaualuma*)" (Holmes, 1957, p. 319, italics added). Pitt (1970) contended that wives "rejected because of infertility joined the *aualuma* . . . partly to instruct the young girls or men on sexual matters" and that the *aualuma* was "freely available to all comers [giving] early travellers an impression of Samoan license" (p. 163).

From these accounts, we can deduce several possible reasons why the missionaries so radically transformed the *aualuma*.

First, because of the missionaries' patriarchal orientation, the experiences and rights of women were less likely to be respected than those of men in the *'aumaga*. As seen earlier, precontact Samoan culture appears to have been less patriarchal than is now the case. Women seem to have had more freedom of self-determination, more social value, and higher status economic roles. Christian influence appears to be the force that began the decline in the status of women there (cf. Lockwood's, 1993, analysis of Tahiti).

Second, the practice by which young women lived together, supported one another emotionally, and functioned sexually in a semiautonomous manner, would have been taken as a threat to both Christian morals and to the *merging* patriarchy of the missionaries and Samoan *matai*. Undoubtedly, the semiautonomous sexuality of the young women would have been too much for the puritanical missionaries to handle. It is little wonder, then, that they instituted the practice that young unmarried women be encouraged to live in the pastor's house under his and his wife's supervision (of course, they were only accepted if they were still virgins, or had been forgiven for their "immorality").

Third, with the demise of the *taupou* tradition, there was also less legitimation for the *aualuma* in terms of an appeal to traditional authority, so through time less social value would have been attached to it by adult male Samoans. In this context, we can place Keesing's (1934) remarks about the resistance to empowerment of the new Women's Committee in the 1920s: "At the same time resistance has been generated among many of the male titleholders against enhancement of women's authority. This forms another of the currently critical areas of conflict in modern Samoan society" (p. 56).

And, finally, because their function was more "expressive" than that of the "instrumental" *'aumaga*, it would have been less disruptive economically to disband the *aualuma*. All these factors would have operated in the opposite direction in terms of maintaining the basic structure of the *'aumaga*. The *'aumaga* did not threaten patriarchal or puritanical sentiments, it was not tied to the *taupou* tradition, and it was necessary for the day-to-day operation of the agrarian economy.

Eventually, however, the *'aumaga* would diminish in importance as well. But its collapse was to take much longer, and has only taken place recently, largely because of the imposition of the wage economy and mass education. It is important to recognize therefore that while the demise of the *aualuma* seems to have been a direct result of missionary efforts, the demise of the *'aumaga* was more likely an *indirect* result of their efforts, to the extent that they laid the basis for and encouraged the wage economy.

Holmes (1957b) provided an account of the *'aumaga* of the 1950s. We can note from his description that, as late as the 1950s, young males were still formally welcomed to both adulthood and to the local polity. Moreover, their activities in the *'aumaga* were still basically the same as they had been for centuries:

> When a boy has completed his schooling he is eligible to enter . . . the *'aumaga*. His family head (matai) may suggest *'aumaga* affiliation or in some cases a boy goes to his family head and expresses his desire to join. If the family head agrees that he should, he goes with the boy, bringing a kava root in his name to the village council. This presentation represents the official recognition by the village council that a young man may enter the untitled men's society.
>
> After being thus recognized by the village council, he prepares for his actual entrance into his society. At their next meeting the boy goes by himself with his gift of food (*momoli*), which usually consists of a six pound tin of corned beef and other food such as taro, breadfruit and . . . *palusami*.

After placing his gift in the center of the floor the boy takes a seat in the back of the house and waits patiently while the business of the group is discussed. Finally he is recognized in a welcoming speech by one of the "Talking Chief's sons" (*tamato'oto'o*). He replies with a speech, after which one of the Talking Chief's sons distributes his food, and he is officially accepted as a member of the group. Thereafter he takes a seat at a post corresponding to the rank of his family head . . . (pp. 408–409)

Just three decades later, however, we find that throughout most of Samoa, the *'aumaga* appears to exist mainly in name only. According to Shore (1982), during the 1970s, the *'aumaga* in the village he studied in Savai'i still functioned to do the bidding of the *fono* and it provided a route to *matai* titles through *tautua*. Its function seems to have been more "socioemotive" than "instrumental," however. For example, although Shore noted that it functioned to encourage "in its members the development of skills and attitudes appropriate for young men" (p. 102), its members did things such as checking taro quotas once a month and cleaning village and school grounds one day a month.

For the most part, then, villages no longer engage in cooperative labor, except in the service of the village pastor (O'Meara, 1990). Instead, individual wage labor has replaced cooperative village labor. Now, most teenaged males (and females) attend secondary school and many move out of the village upon graduation, in search of the wage labor they have been taught to want. Most move either to larger urban areas of Samoa or to other countries (as is seen in chapter 7). Consequently, even many remote villages have been depopulated, particularly of young males (O'Meara, 1990; Yusuf & Peters, 1985). Accordingly, work projects are rarely communal enterprises any longer:

A few young men occasionally form temporary clubs, working each other's plots in turn or hiring out to work on their families' plots. Sometimes the few families who are short on labor but have money to spare also hire a few neighbour youths for a single day of labor. Otherwise, each nuclear family or each unmarried man usually works his own taro plots independently. (O'Meara, 1990, p. 60)

The point to be taken here is that the *'aumaga* has declined in importance as secondary education has transformed both social relations in each Samoan village and the values held by the young attempting to come of age. The nontraditional values send many of the young off on a search for wage labor in pursuit of the material benefits believed to accrue from this labor. There is little sign that this transformation will slow in pace, or undergo a reversal.

Hence, as was the intention of the missionaries (Turner, 1861/1986,

p. 91), one of the most enduring influences on adolescence and the transition to adulthood has been mass education. Originally, the missionaries were concerned only with teaching adults to read well enough so that they could read the bible themselves (during the many prayer times established as part of the daily routine). At the same time, they needed to solidify their own base of operations and ensure their perpetuity. Meleisea (1987a) described how their plan unfolded:

> Education was a major programme of the mission. The policy was to educate men as pastors to take care of the parishes in the villages, so that when every village had a Samoan pastor, the English missionaries could devote themselves to teaching in the church schools and colleges, and take care of the administration of the church.
> The pastors and their wives ran schools for both children and adults in villages throughout the nineteenth century and, up until the 1950s, most Samoans were educated by village pastors. . . . The pastor taught the boys whatever practical skills he had learnt, while his wife taught the girls *papalagi* domestic arts. (pp. 59–60)

When the state assumed support for education in this century, primary schooling became almost universal. The majority of those in their teen years now attend secondary schools in both Western Samoa and American Samoa. Although it is true that mass education has many social and personal benefits, in the case of Western Samoa, it has some very discernibly negative side-effects, given the limitations placed on this society by its geographical and economic circumstances. As Schoeffel and Meleisea (1983) noted, Mead (1928, e.g., pp. 28–29) recorded the first stages of the social disorganization caused by mass education in the 1920s.

Most devastating to Samoan culture is the fact that mass education has led to a decline in respect for ancestral wisdom, and for those who possess it. The respect for the elder that was once a mark of this culture is diminishing as children learn competing sources of wisdom and knowledge. The world views explicitly and implicitly inculcated through education sometimes come into direct conflict with the world view of the *aiga* or *matai*. This can be a source of great difficulty for everyone concerned, especially if the young person must then live under the authority of the *matai* (cf. O'Meara, 1990).

As a consequence of the decline of the importance of the *'aumaga*, the expansion of the secondary school system, and a loss of respect for ancestral customs, many young Samoans have sought out nontraditional sources of meaning. Unfortunately, not many other forms of meaningful activity have become available. As is seen in chapter 7, wage unemployment in Western Samoa is widespread, but it is particularly high among the young. For a time, emigration to New Zealand from Western Samoa was easily accomplished and constituted somewhat of a rite of passage (MacPherson & MacPherson, 1985). However, this avenue has now been largely cut off by the financially beleaguered New Zealand government (Immigration Division, New Zealand, 1982; Yusuf & Peters, 1985). Educated to think beyond the *'aumaga* and *aiga*, but with few viable options, many young Samoan men have felt extreme desperation (cf. Leacock, 1987; MacPherson & MacPherson, 1985; Norton, 1984).

To summarize this chapter thus far, succeeding generations of young Samoans, having been denied traditional rites of passage by the missionaries, were told that they could "prove themselves" as individuals by being successful economically in an occupation or career. They were also told that Western-style, disciplinary, mass education made the person "better," and better able to understand the world to come. Young people were taught to "strive" in this system. But this Western obsession with accomplishment has reduced the value of striving defined in terms of traditional *fa'a Samoa*. Traditionally, men displayed their ambition through hard work within the *'aumaga*, by earning respect through the *tautua*, developing oratory skills, and perhaps earning the title of *matai*. Instead, successive generations have learned to sell their labor outside the *'aumaga*, and to keep the fruits of that labor for themselves and perhaps their *'aiga*. In doing so, many feel they have lost their Samoan identity.

MISSIONARY IMPACT ON SEXUAL PRACTICES

As we have seen, much has changed since the missionaries set out to reform the "heathen immorality" of Samoans. In particular, the rights and privileges of young females have suffered. Collectively, they seem to have gone from having membership rights associated with a group that had solidarity and semiautonomy in the *aualuma*, through to being sequestered by the local pastor to "protect" their chastity, to being peripheral to a nominal "committee." In other words, they have been culturally disenfranchised by changes initiated by missionaries, and their status has decreased significantly in Samoan society as a result. Some of the stages of this transformation can be culled from the writings of missionaries, diplomats, and scholars.

The only relevant missionary account from the present century comes from Brewer (1930/1975), who was posted to Tutuila from 1920 to 1923. We can detect from his record both the missionary influence in altering Samoan public behavior and traces of traditional customs that might have been practised when the missionaries were *not* a part of group activities. For example, he wrote of an experience where he encountered a *malaga* (visiting party) on its way to Pago Pago:

> In the forefront was the taupou . . . She was really prancing along, and *she didn't have a single solitary stitch on her*! She stopped, and *when she saw we were missionaries*, she immediately grabbed her lava lava . . . and threw it around herself, pulling over her shoulders a little shawl that she had been carrying on her hand. (p. 27, italics added)

Also writing at this time was a New Zealand administrator, N. A. Rowe (1930). He related a similar experience from Western Samoa during his stay there during the 1920s:

> Sometimes, going along a road through the dripping forest in Samoa, one comes upon a party of girls, bare-breasted, wearing the sufficient lavalava, and carrying their frocks upon their arms. At the sight of a white man they will stop and struggle into the redundant garments, or hurriedly, as they pass, will hide their breasts in simulated shame. *This is one of the major triumphs in Samoa, of religion.* (p. 47, italics added)

Those familiar with contemporary Samoa will recognize that "sociosexual" conditions have changed considerably. Rowe (1930) appears to have sensed these changes and their deleterious effects, as the following quotation from the Colonial Office Report (1924–1926) suggests. Note in this passage the perception that Christian morals were producing forms of sexual deviance and a moral "decadence" that had not existed in precontact Samoa:

> The spirit of the old severe system is gone; it has been replaced by the mere letter of the new, to which the native accords lip-service without understanding. He is a man deprived of moral landmarks. *Clothes, covering bodies which once went naked and unconscious, have contributed to his moral decadence by stimulating nasty curiosities, which never before existed.* (pp. 48–49, italics added)

We can gain from these selections a sense of the practices that likely prevailed in precontact Samoa and the changes in these practices initiated by missionaries. Yet, during the first 100 years of contact with missionaries, change was apparently not uniform throughout the Samoan islands. Rather, it is quite likely that various combinations and permutations of precontact and imported-Christian practices would have been found in different locations throughout the islands. In the 50 years since, however, a greater homogeneity of practices seems to have been brought about through the concerted efforts of religious authorities, and has been aided by modern mass communication and transportation.

Thus far, little has been reviewed that places Mead's account of the incidence of premarital sexual behavior during the 1920s in doubt.

Moreover, we can even move past the period when Mead made her observations, and gain a glimpse of how some precontact practices survived as far as the mid-1900s. Holmes (1957b) provided the following account of what he found in the village of Ta'u during the 1950s:

> Sex activity in Manu'a centres around the boy's intermediary (*fa'asoa*) who approaches the girl and arranges a rendezvous with the boy. . . . Such meetings take place after the families of the principals have retired, when they slip out of their houses to a specified rendezvous. . . . *Samoan girls enjoy a considerable amount of sexual freedom without the conflict as that confronting the American teen-aged girl.* Promiscuity is condemned by the church but winked at by the family. An unmarried girl who finds herself pregnant will face a certain amount of verbal abuse from her family, but the matter is soon forgotten, and the newborn child is welcomed with open arms, without any stigma attached to it. (pp. 409–411, italics added)

It is unclear to what extent these attitudes and practices have survived intact in the 1990s, but Schoeffel and Meleisea's account suggests they have (see chapter 3).

One area where the missionaries seem to have succeeded fully is with instilling the puritanical obsession with covering the human body—an uncomfortable thing to do in a tropical climate. By the 1960s, many Samoans seemed to have fully embraced Christian attitudes toward the body. For example, based on his experiences as a school teacher in Western Samoa, Irwin (1965) wrote:

> Samoan girls are much more conservative in their dress than European girls. They dress colourfully and attractively but never daringly. Clothes are meant to hide the figure, particularly between the hips and knees. No Samoan girl wears shorts or a bathing costume. If she did, everybody would know she was a bad girl like those who come ashore half naked from passing ships. (p. 54)

This is indeed quite a different Samoa than the one recorded by the early missionaries. Now, the free-spirited, half-naked women are *palagi* ("European") tourists, and the people being offended by the lack of modesty are Samoans. In this respect, it appears that Europeans have traded places with Samoans, in the sense that Samoans have imitated elements of European culture of the Victoria era (e.g., modesty) and Europeans have imitated elements of precontact Samoan culture (e.g., immodesty). It is interesting to note that in the original edition of her book, Mead provides a photograph of a young woman wearing only "the bark costume of long ago" (caption of photo, between pages 160 and 161). This young woman did not appear shy in baring her breasts for the photographer.

Although the photographer is not identified, it was likely Mead herself, and the subject was likely one of her informants.

The Victorian prudishness acquired from Europeans apparently even extends to the clothes worn to school, at least for Samoans now living in New Zealand. The following passage from a monograph prepared for the New Zealand Office of Race Relations (to help New Zealanders understand Samoans) explains this:

> In co-educational schools, embarrassment may be felt if the school insists on certain uniform requirements for girls. Samoan girls do not always feel comfortable in short physical education skirts or in rompers. In particular they are sensitive about wearing swimming costumes in the presence of boys, and if they do, the boys too can feel embarrassed. (Ngan-Woo, 1985, p. 21)

Undoubtedly, we are seeing the "ideal" of the Christian-Samoan culture in that passage. We also see the ideal in the following passage regarding what is claimed to be the "real" sexuality of Samoan women:

> Samoan female adolescents are aware of the taboo placed on their womanhood. The *aiga* take pride in its daughters. Virginity is to be yielded only at marriage. Brothers or male cousins exercise an active surveillance over their women's whereabouts, but will avoid their women's physical presence. They take an active interest in activities entered into by women, especially those at night. . . . The Samoan ideal of chastity for females before marriage is rigidly upheld and it becomes a source of pride for an *aiga*. Thus a son's *triumph in marrying a virgin is seen as especially significant.* (Ngan-Woo, 1985, p. 24, italics added)

That it would be a "triumph" to marry a virgin, if virginity were a universal custom in the culture, is of course a logical contradiction. However, we see something of the *reality* in a passage following shortly after the statement just cited: "In fact however, some Samoans do produce 'illegitimate' children, but the woman is still able to marry. In other words, faaSamoa recognises human frailty, and is willing to forgive" (p. 24).

Based on the material just examined, together with material from chapters 3 and 4, there is little reason to doubt the basic details of Mead's account: We may take exception with how she expressed certain things and how she explained other things, but there is little reason to doubt the likelihood that some of her informants engaged in a certain amount of casual sexual behavior that was passively accepted by their communities. We have seen strong evidence that such behaviors would have been found in precontact Samoa and we have examined the influence that was changing them when Mead conducted her study (the Christianization undertaken by missionaries). But even with the greater impact of Christian influence in contemporary Samoa, such behaviors apparently still occur (Schoeffel & Meleisea, 1983), and have even been documented by Freeman (1983) himself. Accordingly, in the concluding section of the next chapter I attempt to understand the logic by which Freeman could muster a case against Mead on this issue.

CHAPTER SIX

Mead's Samoa

IS IT PLAUSIBLE IN TERMS OF THE HISTORICAL EVIDENCE?

With the historical perspective developed in chapters 4 and 5, we are now in a position to judge whether "Mead's Samoa" of the 1920s can be pinpointed along the trajectory of social change initiated by the missionaries in the 1830s. Using this technique, we have another means by which to assess the plausibility of Mead's coming-of-age thesis. As discussed previously, the controversy over Mead's work has tended to ignore the substance of her book; namely, the coming-of-age thesis regarding the psychological processes and social structures associated with adolescence in 1920s Ta'u. Chapter 4 presented a reconstruction of the institution of adolescence as it might have existed in precontact Samoa, and chapter 5 traced the changes in that institution initiated by missionaries. This chapter provides a review and analysis of what Mead actually wrote about the Samoa she observed.

Qualifications Placed by Mead Upon Her Findings

Before examining Mead's Samoa it is important to gain an appreciation of the limitations that she herself put on the validity and generalizability of her findings.

To be specific, Mead's (1928) *Coming of Age in Samoa* is about 50 young women between the ages of 8 and 20 who were growing up in three small villages on the island of Ta'u in 1925–1926. The subtitle, *A Psycho-*

logical Study of Primitive Youth for Western Civilization, conveys Mead's intention to focus on the psychological world of these young women and to compare it with the world confronting young American women.

Moreover, the book can actually be read as two overlapping books. The first comprises 14 chapters written for the general public. It is partly because of the literary license used by Mead in these chapters that various interpretations and misunderstandings of the book have emerged. Clearly, one has to read these chapters keeping in mind the fact that Mead was trying to "sell" the book and reach a popular audience with a particular message. The "second book" can be found in the appendices, which present much of the technical, social-scientific, as opposed to the popular, record of her research. Quite possibly, her publisher had her squeeze this information into these appendices to make room for more stimulating material. These appendices provide more exacting details of her research as well as a series of qualifiers regarding her analysis and conclusions.

In Appendix II, Mead presented most of the restrictions that she thought should be taken into account in reading her study. But, of course, as Freeman's critique reveals, she did not always make these restrictions explicit when she described her informant's experiences. Had she done so, the controversy over her work either might not have arisen at all in the first place or it might not have been as complex.

Mead dealt first in Appendix II with the problem of the transitional nature of the culture she was studying and how this made it "impossible to present a single and unified picture of the adolescent girl in Samoa." Thus, she grappled with the problem of including

> descriptions of customs which had fallen in partial decay under the impact of western propaganda and foreign example . . . [and the] double necessity of describing not only the present environment and the girl's reaction to it, but also of interpolating occasionally some description of the more rigid cultural milieu of her mother's girlhood, [all of which] mars to some extent the unity of the study. (p. 259)

We can see here the dilemma Mead faced in providing an accurate account of the experiences of her informants while not letting her research shift to a study of social change.[1] This was to be the decision that in-

---

[1] In Appendix III, Mead provided an analysis of the social change that places her study in historical context. Although much of this is helpful and apparently accurate, in reading it we must bear in mind Schoeffel and Meleisea's criticism that Mead sometimes confused the ideal culture with actual behavior. Thus, for example, her citation of traditional harsh punishments for sexual impropriety (p. 273) must be interpreted in light of what we reviewed in chapters 4 and 5. Yes, such punishments were possible because of stipulations originating in the ideal culture, but it is more likely that they were employed infrequently, and that sexual license was more widely enjoyed.

vited the current controversy and it is something that must be kept in mind when reading her passages about the "typical" experiences of her informants.

A second qualifier specified the limitations of her perspective. She acknowledged that given the qualitative nature of her study with its case history method, "[t]he conclusions are also all subject to the limitation of the personal equation. They are the judgements of one individual upon a mass of data, many of the most significant aspects of which can, by their very nature, be known only to herself" (p. 261). In chapter XIII, she also admitted the possibility for error, but attempted to minimize it in her work by focusing on the ways that "aspects of Samoan life[,] which irremediably affect the life of the adolescent girl[,] differ from the forces which influence [American] girls" (p. 198).

A third qualifier was that, because of her focus on young women and the necessity of keeping her analysis simple, she provided details of Samoan culture *only as they were relevant to her young informants*. In her words: "Although knowledge of the entire culture was essential for the accurate evaluation of any particular individual's behaviour, a detailed description will be given only of those aspects of the culture which are immediately relevant to the problem of the adolescent girl" (pp. 262–263).[2]

Thus, Mead's description of Samoan culture in *Coming of Age* was *tailored* to the experiences of her informants, although she certainly did not make this clear throughout her book. It also appears that she may have forgotten this stipulation herself when she wrote about Samoan culture in some of her later works.[3] Decades later, in the 1969 edition of *Social Organization of Manu'a*, she was to repeat this specification when she acknowledged that she likely saw the culture largely through the eyes of her young informants (see chapter 1, this volume).

A final stipulation involved the fact that the transitional culture placed further restrictions on her sample size of "average girls." Important to the Mead–Freeman controversy is the fact that about one third (9 out of 25) of her postpuberty sample were directly under "European" influence

---

[2]In fact, she stated this in a similar fashion in the introduction to her book (pp. 11–12), but the implications of that passage are easy to miss.

[3]Barnouw (1983) noted that "Mead had an unfortunate tendency, of which Freeman takes advantage, to make stronger and broader assertions in later publications than she did in her original study" (p. 428). On the other hand, Barnouw also noted that after he had "finished Freeman's book, [he reread *Coming of Age*]. In this reading it did not seem to [him] that Mead's generalizations were phrased as strongly as [he] had come to expect. The explanation for this discrepancy appears when one examines Freeman's citations. Although the focus of [Freeman's] book is on *Coming of Age in Samoa*, many of the Mead quotations come from later works of hers" (p. 428).

with their residence in boarding schools run by the village pastors.[4] This limited the pool of "traditional" respondents on which many of her generalizations are based, but it also gave her a "rough control group," as she described:

> The existence of the pastor's boarding-school for girls past puberty provided me with a rough control group. These girls were so severely watched that heterosexual activities were impossible; . . . they lived a more ordered and regular life than the girls who remained in households. The ways in which they differed from other girls of the same age and more resembled European girls of the same age follow with surprising accuracy the lines suggested by the specific differences in environment. (pp. 264–265)

In this passage we see that Mead's generalizations about the "passive acceptance" of premarital sexual behavior obviously did not apply to the one third of her sample living in the pastors' houses. However, although virginity and chastity were required of these young women, these restrictions were apparently not the source of "adolescent turmoil," as is seen later.

With these qualifications in mind, we can now examine and interpret what Mead wrote about how her informants experienced their coming of age. Thus, we should bear in mind that: (a) she was describing a culture in transition; (b) she had a limited view of the culture; (c) she tailored her discussion of the culture to those aspects most relevant to the lives of her informants; and (d) within her sample of her postpuberty informants there were actually "two adolescences"—one closer in nature to precontact Samoa, and one corresponding to the pastors' attempt to recreate Victorian-European culture.

Female Adolescence

According to Mead, the process of coming of age was a gradual one, begun very early in life. After weaning, infants of both genders were usually put under the care of a girl aged 6 or 7 (p. 22) who, with the aid of others, essentially mentored the young one. The girl's duties included keeping the younger child out of mischief and "housebreaking" the child, by using both verbal and physical punishments. For girls, this tutelage

---

[4]The pastors of each village boarded teenagers whose "parents wished them to have three to four years of superior educational advantage and stricter supervision" (p. 68). On Tutuila there was a large boarding school for girls and boys, referred to as the Missionary Boarding School. These "trained many boys as native pastors and as missionaries for other islands, and many girls to be pastors' wives" (Mead, 1928, p. 270). Mead also noted that the missionary schools "were regarded as a social rather than a religious adventure" (p. 161).

lasted until they were sufficiently trained and socialized themselves to become nursemaids.

After this time, the young girl learned various skills such as the rudiments of weaving, collecting and breaking coconuts, spreading copra to dry, playing games and singing, house cleaning, preparing pandanus leaves for weaving, and running errands. Partly because of her young charges, and partly because of discouragements from young boys, the younger girl did not learn how to deal with the reef or "the more adventurous forms of work and play" (p.27). For these reasons, the girl also tended not to engage in cooperative activities and therefore did not learn the principles of group organization as well as did boys. On the issue of play for Samoan children, Mead wrote that despite the assignment of work tasks to them ("which have a meaning in the structure of the whole society"), this did not mean "that they have less time for play than American children who are shut up in schools nine to three o'clock every day" (pp. 226–227).

When the young girl became strong enough, she was relieved of her nursemaid duties, and could work on the plantations and carry produce back to the village. This usually occurred just before puberty. According to Mead:

> It may be said with some justice that the worst period of their lives is over. Never again will they be so incessantly at the beck and call of their elders, never again tyrannised over by two-year-old tyrants. All the irritating, detailed routine of housekeeping . . . is . . . performed by children under fourteen years of age. (p. 28)

Released from baby-tending, young women now learned more complicated skills related to the carrying and preparation of food. Then, if they proved themselves capable of completing tasks competently, they could go on fishing expeditions and engage in all of the preparations for these. They could also learn the techniques of harvesting forest products for weaving, clothing and ornaments. The most important skill acquired was weaving the "fine mat," considered "the high point in Samoan weaving virtuosity" (p. 32). (In general, woven mats are very important in daily Samoan life, being used for sitting, sleeping, eating, and privacy, whereas fine mats have a complex exchange value in Samoa culture; see Linnekin, 1991.) These fine mats should only have taken 1 to 2 years to weave, but for the adolescent who was not yet an accomplished weaver, a longer period was allowed. Begun at puberty, these mats were often not finished until age 19 or 20, after which they became part of the girl's dowry. The quality of the mat became "a testimony to the girl's industry and manual skill" (p. 32).

The period during which the fine mat was being woven can be considered a moratorium—a socially permitted delay of adulthood (Erikson, 1968, discussed in chapter 8, this volume; cf. Mageo, 1988). The social permission existed in the sense that the young woman could not marry until the fine mat was completed, because it was part of her dowry. On the one hand, this seems to have been a period of psychological preparation and of proving qualifications of marriageability. On the other hand, this period was the freest of the young woman's life, particularly in being relieved of child-tending responsibilities. In reference to the first characteristic of this period Mead wrote that:

> Throughout this more or less systematic period of education, the girls maintain a very nice balance between the reputation for the necessary minimum knowledge and a virtuosity which would make too heavy demands. A girl's chances of marriage are badly damaged if it gets about the village that she is lazy and inept in domestic tasks. But after these first stages have been completed the girl marks time technically for three to four years. (p. 33)

This "marking of time" involved completing basic required tasks and demonstrating a "pass proficiency" (p. 37) in terms of competence and industry—and therefore marriageability. To excel would end the moratorium period, by inviting marriage and children, and she would be denying herself the opportunity for freedom and experimentation. Mead described this as follows:

> She does the routine weaving, especially of the Venetian blinds and carrying baskets. She helps with the plantation work and the cooking, she weaves a very little on her fine mat. But she thrusts virtuosity away from her as she thrusts away every other sort of responsibility with the invariable comment, *'Laititi a'u'* ("I am but young"). All of her interest is expended on clandestine sex adventures, and she is content to do routine tasks. . . . (p. 33)

Obviously, not all of Mead's informants were expending all of their interest on "clandestine sex adventures," because one third were under the supervision of the pastor, specifically to prevent such behavior. Again, it is important to read Mead with the qualifications just discussed in mind.

In terms of preparation for adulthood, most instrumental task learning had already taken place by this point. However, although marriageable,

> the seventeen-year-old girl does not wish to marry—not yet. It is better to live as a girl with no responsibility, and a rich variety of emotional experience. This is the best period of her life. . . . The long expeditions after fish and food and weaving materials give ample opportunities for rendezvous.

Proficiency would mean more work, more confining work, and earlier marriage, and marriage is the inevitable to be deferred as long as possible. (p. 38)

Readers are reminded that in chapter 4, missionary John Williams (cited in Moyle, 1984) was quoted as writing that "young females I have heard has a great objection to marriage[,] a roving commission being more congenial to their feelings" (p. 233).

Although the account thus far describes a more "traditional" coming of age, the question arises as to what was taking place for those boarding at the pastor's house. One would think that the imposition of European standards of conduct would introduce stress and conflict into the lives of these young women as they struggled with the conflicting requirements of two cultures. Mead dealt extensively with this issue, and concluded that although some conflicts had been experienced by several young women (her "upward deviants" who wanted to move beyond traditional Samoan culture), these conflicts were not a source of sustained personal distress for them. On the other hand, she predicted that this would not be the case in the future, as institutions such as mass education drew young Samoans away from traditional culture. In her words:

> while religion itself offered little field for conflict, the institutions promoted by religion might act as stimuli to new choices and when sufficiently reinforced by other conditions might produce a type of girl who deviated markedly from her companions. That the majority of Samoan girls are still unaffected by these influences and pursue uncritically the traditional mode of life is simply a testimony to the resistance of the native culture, which in its present slightly Europeanized state, is replete with easy solutions for all conflicts; and to the apparent fact that adolescent girls in Samoa do not generate their own conflicts, but require a vigorous stimulus to produce them. (pp. 170–171)

The primary fact of life introduced for those who wished to reside in the pastor's house was the requirement of chastity, something that appears to not have been required of most young women in precontact Samoan culture (chapter 4). But, as Mead wrote, "[t]he lives of the girls who lived in the pastor's household differed from those of their less restricted sisters and cousins only in the fact that they had no love affairs and lived a more regular and ordered existence" (pp. 156–157). If the requirement of chastity was unacceptable to the young women, or if the living arrangement was unpleasant, she simply had to break an important rule of the pastor's household and she would be expelled (p. 163). In fact, Mead reported that almost all of her older informants had spent some time in a pastor's house (p. 161). However, a number of her informants apparently found the experiences in the pastor's house to be pleasant, and life to be more leisurely there than with their families, so they obeyed the rules. In Mead's words, their fascination with the experiences in the pastor's house "was usually a sufficient bribe to good behaviour, at least to discretion" (p. 163). In a sense, then, a balance was struck to maintain the moratorium period of precontact culture by which restrictions on sexual behavior were exchanged for other freedoms and rewards.

As for the missionary attitude toward chastity, Mead contended that a blind eye was turned toward the premarital sexual behavior of those not residing in the pastor's house, as long as it was discrete (as noted earlier, forms of discretion also appear to have been necessary in precontact Samoa). The "passive acceptance . . . of pre-marital irregularities went a long way towards minimizing the girls' sense of guilt" (p. 161) and it went a long way in preventing stress in the lives of the young. As Mead noted, the introduction of Christian values regarding sexuality could "have provided a real setting for conflict" (pp. 163–164). However, as late as the 1920s, the pastors were not enforcing those values, except through residence in the pastor's house. As Mead explained, pastors did not then require church membership or attendance among the young. Had they imposed religious observance on the young, "crises in the lives of the young people would very likely have occurred" (p. 164). Instead,

> the whole religious setting is one of formalism, of compromise, of acceptance of half measure. The great number of native pastors with their peculiar interpretations of Christian teachings have made it impossible to establish the rigour of western Protestantism with its inseparable association of sex offenses and an individual consciousness of sin. And the girls upon whom the religious setting makes no demands, makes no demands upon it. They are content to follow the advice of their elders to defer church membership until they are older. *Laititi a'u. Fia siva* ("For I am young and like to dance"). The church member is forbidden to dance or to witness a large night dance. (p. 164)

Thus, we can see the unique mixture of Samoan traditions and Victorian Protestantism in 1920s American Samoa. Samoan-born pastors were faced with the dilemma of *imposing* "imported" values on their own people. Undoubtedly, the imposition of such values on an unwilling population would have been at great cost or it might have had a "boomerang effect" leading to forms of resistance. As we saw in chapter 5, however, the missionaries had an ambitious, long-term agenda of transforming the entire culture. Accomplishing a goal of this significance takes generations, and must be done at a pace that does not estrange the people being changed. From Mead's record, we see that they chose to transform the culture through the formal education of the young, where they could isolate the young from their families and indoctrinate them with Christian values and European knowledge.

The residential arrangements in the pastor's house appear to have replaced the *aualuma* of precontact culture as a significant force in the socialization of young women. The *aualuma* provided the young woman with a peer group and a meaningful, structured, daily existence, both of which were now rendered by the pastor. Those who did not live in the pastor's house, therefore, were more on their own, and were left more to their devices, as Mead documented. Indeed, Mead attributed some "individual peculiarities" of personality among some of her informants to the absence of a homogenizing effect of a socializing institution like the *aualuma* (p. 140).

In reference to the *aualuma* of 1920s Ta'u, Mead documented its state of decay as an institution. She noted that formerly "[t]he girl's entrance into the *Aualuma* was *always, not just occasionally*, marked by a feast" (p. 275, italics added). By the time of her study "the formal entry into the *Aualuma* is often neglected and is more a formal fee to the community than a recognition of the girl herself" (p. 80). When formal entry was observed, it was often not until 2 to 3 years past the girl's puberty. Even then it might simply involve her *matai* sending "an offering of food to the house of the chief *taupo* of the village, thus announcing that he wishes the daughter of his house to be henceforth counted as one of the group of young girls who form her court" (pp. 76–77). Of course, with the *taupou* tradition obsolete, there was little concrete justification for maintaining the *aualuma*. In fact, when Mead was there, in "many parts of Samoa[,] the *Aualuma* has fallen entirely to pieces and is only remembered in the greeting words that fall from the lips of a stranger" (p. 77).

This account constitutes the outline of the institution of adolescence available to Mead's informants. With respect to entry into adulthood, the process was straightforward—find a suitable mate, or have one arranged, and get married. This generally happened at age 20 or 21. If it did not happen voluntarily at this age, social pressures were brought to bear upon the woman, and her attitude changed:

> The girl of twenty-two or twenty-three who is still unmarried loses her laissez faire attitude. Family pressure is an effective cause in bringing about this change. She is an adult, as able as her married sisters and her brother's younger wives; she is expected to contribute as heavily as they to household undertakings. (p. 185)

Therefore, it was the need for the young woman's domestic labor that in part forced an end to her moratorium period and pushed her to marry. Therefore, she turned her attention to marriage and "settled down to increase her value as a wife" (p. 185). Once married, life was highly structured, but it was apparently generally satisfying, and Mead believed "[t]he young married women of twenty to thirty [to be] a busy, cheerful group" (p. 190). Mead noted an exception to this with Manita, who was 27 and still unmarried. Manita was mentioned as an example of an "upward deviant," although she was not in Mead's sample of 50 informants. Manita was considered the most beautiful woman in the village and had many "suitors and lovers," but "she was of a haughty and aggressive nature and men whom she deemed worthy of her hand were wary of her sophisticated domineering manner" (p. 166).

Based on the observations just outlined, Mead concluded that most of her informants experienced their adolescence to be relatively unstressful. Mead attributed this to the "general casualness of the whole society" (p. 198) and to the "primitive" or uncomplicated nature of the society. In interpreting her conclusion, however, we must bear in mind the qualifications placed on her study. For example, she was studying a culture in transition, so her conclusions apply more to the traditional structures and customs, than to the Western customs being introduced by the missionaries.

With this in mind, we can more readily accept her conclusion that the remnants of precontact Samoan society provided a relatively straightforward socialization process for the young as they came of age. Contributing to this would have been an absence of competing belief systems as well as a lack of ambiguity regarding the roles that constituted the adult identity (both of which the missionaries were chipping away at, however, as Mead recorded). With little to produce stress and confusion in making choices, the young would come of age in a community where the practices and beliefs in one family or village were very much like the practices in other families or villages. All of this, along with the Samoan extended family (large numbers of siblings and relatives to diffuse emotions and tensions), appears to constitute a formula for an adolescence uncomplicated by psychological distress or an acting out of conflicts. There was simply very little that would produce much "emotional turmoil" to get in the way of individuals enjoying their moratorium and then accepting their adult roles.

Mead mentioned sexual freedoms, of which the young could take advantage, in the context of the ease of passage through adolescence. Although it does appear that a number did exercise this franchise, a significant proportion did not, yet according to Mead they too experienced an ease of passage. Logically, then, sexual freedom or the potential for experimentation would not have been crucial to an ease of passage. In fact, Mead did not include sexual freedom as a requirement for an ease of passage. Still, precontact Samoan culture appears to have had a more consistent sexuality morality for both men and women than the morality being introduced by the missionaries. Traditionally, sex was viewed to be a "natural, pleasurable thing" (p. 201), and women were not stigmatized for playing their natural role in it. However, it was also the case that they and their partners had to be discrete about it. In addition, their activities could not be seen as a threat to community interests. Accordingly, these young women should not have been confused or traumatized over conflicting standards for sexual behavior. Because of the controversial nature of these assertions, this issue is examined in more detail here.

Male Adolescence

Mead described female adolescence as constituting a predictable set of steps delivering the individual relatively painlessly from childhood into adulthood. Although her study was specifically about females, she also provided considerable detail about the coming of age of males. But, because her informants were female, she did not have access to the psychological world of the young male. Rather, she had to rely more on her observations of the social structures to which they were conforming, and she had to draw inferences about their experiences from these observations. In contrast to the changes affecting female adolescence, male adolescence had been less drastically transformed by the missionaries. This appears to have been the case because (a) it stood in contravention of fewer Victorian morals, and (b) it was more instrumental to the material well-being of the community. However, we see from Mead's account that the passage into adulthood was often more dramatic and more structured for young men, presumably as a precursor for their more dominant roles in that society.

Early childhood was very similar for boys and girls. By mid-childhood, however, differential opportunities emerged, as Mead described:

> Very small boys also have some care of the younger children, but at eight or nine years of age they are usually relieved of it. Whatever rough edges have not been smoothed off by this responsibility for younger children are worn off by their contact with older boys. For little boys are admitted to interesting and important activities only so long as their behaviour is circumspect and helpful. Where small girls are brusquely pushed aside, small boys will be patiently tolerated and they become adept at making themselves useful. (p. 28)

These young boys would form themselves into groups of four or five and place themselves under the supervision of a group of older males in order to learn various skills, such as reef fishing. These experiences would have also taught them how to quickly organize themselves into cooperative work groups. The learning of cooperative work habits was an important socialization influence in preparing the young male for the requirements for "success" in this culture. These cooperative work habits were important qualities that would eventually carry young males into adulthood. In contrast, as mentioned earlier, with the demise of the *aualuma*, young girls were no longer exposed to these types of group-bonding experiences. This deficit undoubtedly made it more difficult for women to achieve a role of high status in the culture (women could earn titles, but it was rare). During their middle childhood, the boys also learned how to prepare food and cook it in the *umu* (an outdoor oven made by covering food with hot rocks and banana leaves). Males were expected to perform this task for the duration of their lives, so long as they remained "untitled" (i.e., not a *matai*).

Tattooing, the principal rite of passage of precontact culture, was prohibited of young males by the missionaries while Mead was on Ta'u. Thus, Mead (1928) noted that "tattooing has been taboo on Manu'a for two generations, so only part of the population have made the necessary journey to another island in search of a tattooer" (p. 267). She also observed that "today, scarcely half of the young men are tattooed" (p. 275). As we saw in chapter 4, in precontact Samoa, tattooing took several months; it was done in mutually supporting groups; and it was a prerequisite for entry into the *'aumaga*.

With tattooing banned at the time of her study, Mead did not specify whether there were any formal prerequisites for entry into the *'aumaga*. (Holmes, 1987, noted that in the 1950s, finishing school was the entry requirement.) She did note, however, that:

> When the boy is old enough to enter the *'aumaga*, the head of his household either sends a present of food to the group, announcing the addition of the boy to their group, or takes him to a house where they are meeting and lays down a great kava root as a present. Henceforth the boy is a member of a group which is almost constantly together. (p. 76)

Although tattooing was not part of their rite of passage, young men were apparently prepared in terms of skill acquisition for early adulthood. According to Mead:

> the seventeen-year-old boy is not left passively to his own device. He has learned the rudiments of fishing, he can take a dug-out canoe over the reef safely, or manage the stern paddle in a bonito boat. He can plant taro or transplant cocoanut, husk cocoanuts on a stake and cut the meat out with a deft turn of the knife. Now at seventeen or eighteen he is thrust into the *'aumaga*, the society of the young men and the older men

without titles, the group that is called . . . "the strength of the village." (pp. 33–34)

Thus, the experiences in the *'aumaga* were important for socializing the young man to develop the skills and responsibilities associated with adulthood. In addition to a sense of competency and accomplishment, he also derived a sense of belonging in his community and a realization that his family and village needed him. Mead described these socializing experiences, and how they differed from the experiences of young women:

> Here he is badgered into efficiency by rivalry, precept and example. The older chiefs who supervise the activities of the *'aumaga* gaze equally sternly upon any backslidings and upon any undue precocity. The prestige of his group is ever being called into account by the *'aumaga* of the neighbouring villages. His fellows ridicule and persecute the boy who fails to appear when any group activity is on foot, whether work for the village on the plantation, or fishing, or cooking for the chiefs, or play in the form of a ceremonial call upon some visiting maiden. Furthermore, the [male] youth is given much more stimulus to learn [than are females] and also a greater variety of occupations are open to him. (p. 34)

From these accounts, it is clear that males underwent experiences that were more group structured than did females and that they had more choice about what they would learn and do as an "occupation." Occupational choices included house building, fishing, carving, and oratory. Important to the coming-of-age issue is Mead's inference that the period following entry into the *'aumaga* could constitute a moratorium, if the individual so chose. As was the case for females, a young man could make this time a moratorium by not excelling too much, for to do so would make him eligible for higher status, marriage, and perhaps a chiefly title. Accordingly, the young man could "bide his time," carrying out his responsibility but not outdoing his peers.

Thus, membership in the *'aumaga* marked the young man as "grown up and ready to assume adult responsibilities" (p. 55). Although doing his share of the work was important, he should not excel in his duties unless he was ready to accept a title. Once a title was offered, if it was refused, it would never be offered again (p. 36). Consequently, the young man would squander his ultimate chances for advancement. When a title was accepted, however, the freedoms of the moratorium were relinquished and the titled person was expected to become very serious and circumspect, and not engage in open hedonistic activities. Mead did not identify any cases where these pressures caused serious problems of the storm and stress variety for young men. However, she did identify a 27-year-old *matai* who described the burden it had been for him being a *matai* for the past 4 years. Moreover, she noted that "the boy is faced by a far more difficult dilemma than the girl" (p. 37) in trying, on the one hand, to keep the freedom of youth while, on the other hand, meeting his obligations. For he needed to maintain a reputation that would later allow him to find a good marriage partner and to earn a title.

The *'aumaga* also appears to have provided a very effective socialization setting within which to learn the requirements of the formal leadership roles in the community, particularly those of the *fono* (village council), with its esoteric rituals and language. Mead described this in detail: "The *'aumaga* mirrors the organization of the older men. Here the young men learn to make speeches, to conduct themselves with gravity and decorum, to serve and drink the kava, to plan and execute group enterprises" (p. 76).

According to Mead, the practice of entering all young men into the *'aumaga* had the effect of creating a population that was homogeneous in behavior and outlook. As Mead noted, "the many years' apprenticeship in the *'aumaga* formed an excellent school for disciplining individual peculiarities" (p. 140). In terms of the sanctioning of behavior, normative pressures were directed against "the inept, the impudent, and the disobedient among the young" and they encouraged the young to "work industriously and skilfully, not [to] be presuming, [to] marry discreetly, [to] be loyal to their relatives, not [to] carry tales, not [to] be trouble makers" (p. 130).

This discussion provides the essentials of Mead's description of the coming of age for young males. Readers can draw their own conclusions as to whether it was experienced as pleasant for the typical Samoan male. However, this well-structured adolescence and young adulthood is basically the "coming of age" that would have been experienced for numerous generations prior to Mead's study (except the schooling requirements and the prohibition on tattooing). The young men did appear to face a certain amount of competition, rivalry, and bravado, but whether these were experienced by them as psychologically unpleasant or stressful is not clear. Mead did not record any incidents of serious delinquency or personality disorders among the young men, but she did not intensively interview any young men either. What is clear, however, is the crucial point that their coming of age was structured and orderly, and they therefore should have been prepared cognitively and emotionally as children for their adolescence; as adolescents they should have been prepared cognitively and emotionally for their youth; and as youth they should have been prepared cognitively and emotionally for their adulthood.

Comparing Mead's Account of Sexual Practices With Freeman's Criticisms

Freeman (1983) claimed that "Miss Mead" (p. 226)[5] depicted Samoans "as a people for whom free lovemaking is 'expected' among adolescent girls" (p. 240). On the contrary, Mead actually wrote that "sex activity is never urged upon the young people" (p. 232). Disregarding Mead's actual account, then, Freeman argued that

> this conclusion is indeed so preposterously at variance with the realities of Samoan life that a special explanation is called for; as I shall discuss further . . . all the indications are that the young Margaret Mead was, as a kind of joke, deliberately misled by her adolescent informants. (p. 240)

Freeman rightly noted that Mead identified about one half her sample of postpuberty females to be virgins (Freeman, 1983, p. 238). He also argued that "the furtive sexual liaisons in which a minor proportion of adolescent females became . . . involved were recognized by all concerned as shameful departures from the well-defined ideal of chastity" (p. 240). He applied this latter statement to his own research in the 1960s in which he identified 27% of a sample of teenaged Samoan females as nonvirgin, but implied that the statement applies to Mead's sample from the 1920s.

Given what we have already examined in chapters 3, 4, and 5, one may ask whether it is worth considering Freeman's critique any further on the matter of premarital sexual behavior. I believe it is, because it is on this issue that much of the mythology about Mead's book is centered, and it is on this issue that "Samoans are perturbed by Mead's depiction of them" (Freeman, 1983, p. 240). So let us backtrack a bit before examining *exactly* what Mead wrote in *Coming of Age* about this issue.

As discussed in chapter 3, Schoeffel and Meleisea found Mead's descriptions of the sexual behavior of her informants to be quite plausible, even in terms of their own recent experiences in Western Samoa. The prevalence of such behavior in precontact Samoa was verified in chapter 4 from the records of missionaries, and in chapter 5 we saw how the missionaries set out to systematically eradicate this semi-autonomous sexuality. Mead's account therefore accords well historically, and she appears to have described a transitional culture at the point where the pastors had not yet fully enforced Victorian standards on all of the young women.

How, then, do we account for Freeman's vehemence regarding this issue? The reader will recall that Freeman has held an honorary chiefly

---

[5]Mead was actually married at the time. As noted earlier, Freeman patronized Mead in this manner throughout his book.

title since the 1940s (Freeman, 1983, p. xiv). Schoeffel and Meleisea attributed Freeman's disagreement with Mead partly to the fact that his "chiefly rank imposed an insurmountably difficult impediment to his direct knowledge of the private experiences and informal behaviour of young adolescent females." This is because interaction "between persons of different ranks and statuses, particularly those structurally 'opposed' in Samoan culture, is very restricted and formal" (p. 67).

Testimony from another Samoan supports Schoeffel and Meleisea's view. A. Wendt (1983) argued that "[i]n our culture it has always been taboo to discuss sexual matters publicly or freely or in mixed company. The Church's strict Victorianism merely reinforced this" (p. 13). In addition, Wendt clearly believes that Freeman erred in his argument regarding sexual mores, as is evident in the following passage:

> he is correct in stating that we place a great priority on female virginity, we institutionalize it in the taupou, we forbid pre-marital and extra-marital sex and promiscuity. *But it does not mean that a lot of all these various types of sex do not go on.* As in other societies, what we preach publicly to be our standard or morality does not necessarily reflect how we actually behave. There are a great deal of double standards in all societies. In this instance, *Freeman accepts too readily the evidence of our elders, the guardians of our public morality, who, if they are really human, often break, clandestinely, the very standards they set up.* . . . Even in the 1920s Samoa wasn't anywhere near "all virginity," as it were. (p. 14, italics added)

Wendt explained Freeman's error in the following way: "The easily discernible flaws in Freeman's book stem from its polemical form. To prove Mead wrong some of his claims tend towards exaggeration and idealisation. (This idealisation is also perhaps the result of his profound trust in us)" (p. 14).

It is most curious that Freeman's (1983) own evidence clearly shows that virginity was not strictly enforced even in the 1960s. Only 73% of his sample of 43 teenagers were virgins, a figure that is consistent with the historical trajectory of the social change we have traced in the last three chapters. Moreover, among 19-year-olds, only 40% of his sample were still virgins—hardly representative of a well-enforced norm. Freeman (1991) inadvertently acknowledged that his research supports Mead's findings on the issue of premarital sex, when he wrote that "Mead's own data . . . 'reveal' that 52% of her sample of 25 [postpubertal] girls were virgins, *a conclusion not significantly differently from my own*" (p. 120, italics added). He appears to have missed the implications of such a statement (i.e., that 48% were not virgins) because he simply went on to claim that Mead's data were "*quite at odds* with her misinformed statements about the prevalence in Samoa of premarital promiscuity" (p. 120).

The most obvious conclusion one can draw on the matter is that Freeman must be blind to the evidence on this issue, and its implications, because of the "mission" he has undertaken to rescue Samoan honor, as defined for him by his Samoan associates who entreated him to "correct [Mead's] mistaken depiction" of their culture and ethos (Freeman, 1983, p. xv). If Freeman's representation of Mead's account is erroneous, what then did she write? First, let us see how she gauged the impact of missionaries at the time.

Mead said that "there is a definite inverse correlation between residence at home and chastity" (p. 147), although she did not conduct any statistical tests on her data. When we turn to Appendix V of her book (p. 285) we find a table of figures that allows us to evaluate this statement. From the information provided in this table it is possible to conduct these statistical tests and to use residence in the pastor's household as an operationalization of Christian influence on traditional premarital behavior. I conducted these tests and have provided the results in Tables 6.1 and 6.2. In these procedures, residence in a pastor's household is the independent variable and heterosexual experience is the dependent variable.

In Table 6.1 we can see the statistically significant association (Phi = .37) between these variables in terms of the pattern of cell frequencies. Most obvious is the finding that 10 (62.5%) of those not living with the pastor had heterosexual experience, whereas 7 (77.8%) of those living

TABLE 6.1
An Examination of the Influence of
Christianity on Premarital Sexual Behavior:
Mead's 25 Postpuberty Informants

Count Column Percentage Total Percentage Adjusted Residual		Residence in Pastor's Household	
		Yes	No
Heterosexual experience	Yes	2 22.2% 8% -1.9	10 62.5% 40% +1.9
	No	7 77.8% 28% +1.9	6 37.5% 24% -1.9

Chi-Square:
 Pearson = 3.74, $p < .05$.
 Fisher's Exact Test, $p = .06$.
Phi = .37, $p < .05$.

TABLE 6.2
An Examination of the Influence of
Christianity on Premarital Sexual Behavior:
Mead's Informants 1 Year or More Past Puberty

Count Column Percentage Total Percentage Adjusted Residual		Residence in Pastor's Household	
		Yes	No
Heterosexual experience	Yes	1 16.7% 5.6% −2.7	10 83.3% 55.6% +2.7
	No	5 83.3% 27.8% +2.7	2 16.7% 11.1% −2.7

Chi-Square:
Pearson = 7.48, $p < .01$.
Fisher's Exact Test, $p = .01$.
Phi = .64, $p < .01$.

with the pastor had no heterosexual experience. On the one hand, one might be inclined to dismiss these figures because chastity was a *requirement* for boarding with the pastor. On the other hand, if we look for a norm of sexual abstinence from the community, we do not find a strong one, because only 6 (37.5%) of those not boarding with the pastor had no heterosexual experience. Sociologically speaking, then, the fact that most of those not boarding with pastor had sexual experience suggests that the pastors, not the community, set the norm of sexual abstinence.

It is possible to examine the community norm regarding premarital sex more directly if we restrict the analysis to informants who were 1 year or more beyond puberty (defined by Mead as time since first menstruation). This is justified with Mead's statement that puberty was not the "immediate forerunner of sex experience. Perhaps a year, two or even three years would pass before a girl's shyness would relax, or her figure appeal to the roving eye of some older boy" (p. 146). Accordingly, in this analysis the variable "heterosexual experience" is less affected by biological factors.

When we conduct this analysis, the statistical relationship becomes much stronger, with the measure of association (Phi) jumping from .37 to .64 (Table 6.2). In this more biologically mature sample, 83.3% of those *not* living with the pastor had heterosexual experience, whereas 83.3% of those living with the pastor had no heterosexual experience. If we look for a norm of sexual abstinence from the community for this sample, we do not find one because only 16.7% of those living in the community (i.e., not boarding with the pastor) had no heterosexual experience. Indeed, in this case the *behavioral norm* is clearly with heterosexual experience, as Mead's generalizations suggest.

We can interpret these statistics to reflect the attempt by religious authorities to institute the norm of premarital chastity. They also suggest that chastity was *not the community norm*, especially for those 1 year or more past puberty. In other words, virginity was enforced mainly by the pastor, not by the community.

But what exactly did Mead write in deciphering these behaviors? Well, *nowhere* did Mead contend that the Samoans are "a people for whom free lovemaking is 'expected' among adolescent girls," as Freeman claimed (p. 240). Instead, she said that "free experimentation [was] *permitted*" (p. 150, italics added). In addition, she noted that this "permission" applied to the "average" or "plebian young people" she studied, but that the *taupou* was "excepted" (i.e., excluded from) this "free and easy experimentation." In fact, she wrote that virginity "was a legal requirement for" the *taupou* (p. 98).[6] Quite clearly, the statistical analyses just cited support her contention that the typical "plebian" did engage in premarital sexual behavior.

On the other hand, there are several passages from *Coming of Age* that are questionable. As we review these, it becomes clear that Mead must shoulder some of the blame with respect to the mythology now surrounding the sexual content of her book (cf. Feinberg, 1988). Among her references to her informants' sex lives, we see Mead's tendency to embellish her account with flowery prose, presumably to maintain the interest of her nonacademic audience. From an academic point of view, her descriptions sometimes amount to overstatements and it is clear from her other references to sexual practices that not all of her informants did all of the things she described. Indeed, her use of anecdotes often create the impression that certain practices were more common than was likely the case. Among her more embellished passages are the following:

As the dawn begins . . . lovers slip home from trysts beneath the palm trees. (p. 14)

. . . then at last there is only the mellow thunder of the reef and the whisper of lovers, as the village rests until dawn. (p. 19)[7]

---

[6]Mead's sense about the complexity of Samoan sexual mores is evident with her remark that the Samoan "attitude towards virginity is a curious one" (p. 98).

[7]Undoubtedly, these first two incidents happened, but it is unlikely that they occurred on a nightly basis.

> To live as a girl with many lovers as long as possible and then marry in one's village, near one's own relatives and to have many children, these were uniform and satisfying ambitions. (p. 157)[8]
>
> Familiarity with sex, and the recognition of a need of a technique to deal with sex as an art, have produced a scheme of personal relations in which there are no neurotic pictures, no frigidity, no impotence, except as a result of severe illness, and the capacity for intercourse only once in a night is counted as senility. (p. 151)[9]
>
> Samoans rate romantic love and fidelity in terms of days or weeks at most, and are inclined to scoff at tales of life-long devotion. (p. 155)[10]

But in her book we can also find credible descriptions and analyses of the Samoan sexual ethos. For example, she submitted that for Samoans

> sex is a natural and pleasurable thing; the freedom with which it may be indulged in is *limited by* just one consideration, *social status*. Chiefs' daughters and chiefs' wives should indulge in no extra-marital experiments. Responsible adults, heads of households and mothers of families should have too many important matters on hand to leave them much time for casual amorous adventures. *Everyone in the community agrees about the matter*, the only dissenters are the missionaries who dissent so vainly that their protests are unimportant. *But as soon as sufficient sentiment gathers about the missionary attitude with its European standard of sex behaviour, the need for choice, the forerunner of conflict, will enter in Samoa society.* (pp. 201–202, italics added)

We can see in this passage Mead's record of the resistance of Samoans in relinquishing a more pleasurable aspect of their traditional culture, as well as her prediction that the missionaries would eventually prevail. From the testimony of Samoan scholars presented earlier, it appears that the missionaries have still been only partially successful in altering *actual* behavior. Nevertheless, they seem to have been more effective in producing a consensus about the *ideals* of continence.

---

[8] These are many ambitions to place together in one generalization. Clearly, they did not all apply to those in the pastor's house.

[9] Obviously Mead was not qualified to make such clinical judgments, especially epidemiological conclusions regarding all of the inhabitants of the villages she studied. She probably picked up the notions of technique and stamina from boasting (male?) informants and took liberties in drawing the clinical conclusions.

[10] This statement was prefaced with the observation that she found "only one example of prolonged and intense passion . . . in the three villages" she studied (p. 155).

Mead also appears to have understood the way in which sexual relations were socially organized, and she classified these relations as follows:

> Besides formal marriage there are only two types of sex relations which receive any formal recognition from the community—[a] love affairs between unmarried young people . . . whether leading to marriage or merely a passing diversion; and [b] adultery.
>
> Between the unmarried there are three forms of relationship: [a] the clandestine encounter, "under the palm trees," [b] the published elopement, *Avaga*, and [c] the ceremonious courtship in which the boy "sits before the girl" . . . (p. 89)

This description of the first two forms of sex relations between the unmarried reveals that those who engaged in them faced several obstacles that did not facilitate the "free love" atmosphere she ostensibly promoted. Instead, it appears that she was describing the freedom more from the perspective of some of her informants, than from a community consensus that simply called for people to "look the other way" (cf. Schoeffel & Meleisea, 1983). It appears that the clandestine encounters had to be just that—done in secret—and they were restricted to "those of humbler birth" (pp. 99–100). In describing how the young woman dealt with these circumstances, Mead remarked that all of "the adult and near-adult world is hostile, spying upon her love affairs in its more circumspect sophistication, supremely not to be trusted. No one is to be trusted who is not immediately engaged in *similarly hazardous* adventures" (p. 68, italics added). Moreover, children apparently took sport in trying to catch young lovers in the act "under the palms" (pp. 135–136). The hazards presumably were associated with gossip and gaining a reputation as "promiscuous" (e.g., p. 181). Nevertheless, an apparently elaborate system existed by which a *soa* (a friend—a confidant and "ambassador"; pp. 89–90) arranged for two would-be lovers to meet surreptitiously.

The second form of sex relation was apparently less common, as Mead described:

> Elopements are much less frequent than the clandestine love affairs because the girl takes far more risk. She publicly renounces her often nominal claims to virginity; she embroils herself with her family, who in former times, and occasionally even today, would beat her soundly and shave off her hair. Nine times out of ten, her lover's only motive is vanity and display . . . (pp. 103–104)

Again, in both of these forms of sex relations, it does not appear that anything approaching a free love ethos prevailed. Those who advocate the mythology that Mead's book describes such an ethos would do well to give *Coming of Age* a careful reading.

We now have a much better picture of what Mead actually wrote in *Coming of Age* about the sexual behavior of her informants. Freeman seems to be inaccurate in much of his representation of what she wrote. Moreover, the mythology that this is simply a book about adolescent sexuality is just that—fiction. Although Mead discussed intimate details of her informants' lives, there is far more to the book than this, as we have seen in her description of the coming of age of young Samoans.

To return to the main concern in this book, Mead's coming-of-age thesis emerges from the examination given here as entirely plausible. It has not been seriously contradicted by any evidence, notwithstanding Freeman's often selective use of evidence and recurrent misrepresentation of her position. Indeed, "Mead's Samoa" fits with a trajectory of social change that one would expect for the island of Ta'u, as it came under increasing missionary influence. Moreover, on the basis of the evidence examined in this book, Mead's record of the experiences of her young female informants stands as valid in its basic details.

CHAPTER SEVEN

Coming of Age in Contemporary Samoa

WESTERN INFLUENCE AND THE CULTURAL DISENFRANCHISEMENT OF SAMOAN YOUTH

What has been forgotten in the furor surrounding Freeman's critique of Mead's study is the plight of young Samoans currently struggling to come of age. As we have seen, the young people of Samoa have lost most of the institutions that once guided them on a sure path from childhood to adulthood. The degeneration of these institutions began when the missionaries took it upon themselves to drive "Satan" from the islands. Inadvertently acting as the "shocktroops of capitalism," many missionaries set out to instill values and attitudes conducive with wage labor and capital accumulation.

The changes they set in motion have been accelerating in latter part of the 20th century with the entrenchment of mass consumption, mass production, mass education, and mass communications. Moreover, mass transportation has removed protective barriers from Samoan culture, both physical and psychological, so the basic conception of the realm of possibility has been drastically altered for the young Samoan. Most certainly, missionary influences are not responsible for all that has happened in the past 150 years to young Samoans, but the part they played in setting the stage is significant nonetheless.[1]

---

[1] See Oliver (1961) for a broader, political-economy discussion of historical influences in Western Samoa. Gailey (1987) provided a similar discussion of Tonga, bringing out gender issues, as did Lockwood (1993) in reference to Tahiti.

In a sense, the crux of the current "youth problem" in Samoa is that although the *conception* of the realm of possibility has changed for many young people in Samoa, the *reality* of what is possible has not changed significantly. What emerged from the sociohistorical analysis presented earlier is that at the heart of these changes is a "cultural disenfranchisement" of young people by which they have been robbed of their rich cultural heritage, and given poor "Western" economic prospects in return.

For example, the propitious institution that once saw to the coming of age of young women (the *aualuma*) was stripped of its mandate soon after the missionaries were securely established, probably because it offended missionaries' puritanical and patriarchal values regarding the autonomy of women—sexual and otherwise. Young women of contemporary Samoa are not only denied this aspect of their heritage, they also face a denial that there was anything different for Samoan women before contact with the West. Thus, they have lost part of their cultural heritage and been misled about their loss. As is seen in this chapter, there is now a "crisis" in gender roles in contemporary Samoa associated with a loss in the status of women introduced by Western religious and economic influences. Interestingly, it is this turn of events in which Mead's work has been caught, because Freeman took up the cause of a denial of Samoa's non-Christian past.

In addition, changes initiated by the missionaries that set the stage for the introduction of the wage economy eventually led to the disintegration of the auspicious institution that saw to the coming of age of young men (the *'aumaga*). In 1928, Mead predicted that should the *'aumaga* ever be disbanded "Samoan village life would have to be entirely reorganized," because the entire village depends on their work and support of ceremonies (p. 76). Moreover, Mead (1928) predicted that Western influences would be devastating for Samoan culture:

> economic instability, poverty, the wage system, the separation of the worker from his land and his tools, modern warfare, industrial disease, the abolition of leisure, the irksomeness of bureaucratic government—these have not yet invaded an island without resources worth exploiting. (pp. 276–277)

It is difficult to say what the ultimate outcome of these changes will be, but it is hard to be optimistic about how well things will turn out in the short run. What is clear, as is seen later, is that many young people in contemporary Samoa no longer have the benefit of the benign transition from childhood to adulthood described by Mead. The structures that have replaced the traditional coming-of-age institutions leave much to be desired in terms of guiding the young from childhood to adulthood.

Unfortunately, along with other Western practices, Samoan society has increasingly adopted the practice of making childhood and adolescence into a period of emotional and economic dependency. In doing so, many of the basic rights and privileges of its youngest people have been taken away. Meanwhile, many adults themselves have become dependent on Western economic, social, political, and religious institutions. Consequently, the *quid pro quo* (reciprocal exchange) that existed between young people and adults in precontact culture has been upset. Indeed, as we see, young people have lost much in the span of a few generations, particularly much of the cultural heritage that once saw to their basic emotional and material needs. What they have lost has been replaced for many with unfulfillable dreams imported through Western institutions.

Analysis is restricted in this chapter to the situation currently facing young people in Western Samoa. It is necessary to consider Western Samoa separate from American Samoa because the two countries are distinct political entities. As such, they have experienced different economic and political forces over the century. The islands of Samoa were divided between the colonial powers of Germany and the United States in 1900. Upolu and Savai'i (now Western Samoa) went under German rule, whereas Tutuila and the Manu'a Group went under American rule. The latter constitute American Samoa, still a U.S. territory, while the former passed under New Zealand rule during World War I, and became an independent state in 1962.

In terms of the extent of Western influence, although no region is untouched, in Western Samoa, the island of Savai'i has been less affected by the modern wage economy than has the island of Upolu. The greatest Western influence on Upolu is in Apia area, the capital city (and only city) with a population of 45,000. American Samoa has been highly "Americanized," especially around Pago Pago harbor, the capital area—there are no cities per se in American Samoa, rather parts of the island of Tutuila are becoming "suburban." More remote villages and islands in the Manu'a Islands (comprising Ofu, Olesega, and Ta'u) retain some of their traditional lifestyle, but less so than in more remote areas of Western Samoa.

The common thread of their joint Samoan heritage remains, but at present they are on two distinct trajectories into the future. In many respects, those coming of age in American Samoa are far better off materially than their "poorer cousins" in Western Samoa. There are more educational opportunities, including free primary and secondary education as well as opportunities to attend U.S. universities. Because all American Samoans are also American nationals, there is the option of emigrating to the United States, especially Hawaii or California. In addition, many males join the U.S. armed forces (cf. North, 1991). In contrast, Western

Samoa ranks close to Bangladesh in terms of gross national product (GNP), although its largely intact subsistence economy makes the comparison invidious (Meleisea & Meleisea, 1980).

In terms of social problems, there does not appear to be an unusually high suicide rate (King, 1981) or crime rate in American Samoa. Among those who emigrate, some develop problems associated with living in U.S. inner cities, and those who have returned recently have brought back the gang mentality found among some minority youths in the United States. Consequently, there is some current concern on Tutuila about an increase in violence associated with this gang mentality. Incidentally, some of the gang mentality is now finding its way to Western Samoa via American Samoa.

In view of these differences, I am not dealing in detail with the problems of coming of age in contemporary American Samoa because these problems now very much resemble those found among the young in many parts of the United States. Coming of age in American Samoa certainly has changed dramatically since the time of Mead's study, and consequently any research conducted there now would not shed much light on the controversy over her work. But, in choosing between which "coming of age" to discuss in this book, I have selected the situation in Western Samoa because the problems facing the young there appear more pressing than those facing the young in American Samoa. For a treatment of the situation in contemporary American Samoa, see Holmes and Holmes (1992).

COMING OF AGE IN WESTERN SAMOA, 1990

One hundred and sixty years of contact with the West has dramatically altered daily life in Western Samoa, even in remote villages. As was seen earlier, the missionaries set out beginning in the 1830s to change the values and attitudes associated with fundamental notions like individualism versus collectivism, and profit versus sharing. Then, with some difficulty, entrepreneurs introduced practices such as wage labor, making way for capital accumulation (see Oliver, 1961, for Samoan resistance to these influences). As we also saw, generational replacement has facilitated the process of social change, for it was by gradually affecting each subsequent generation that the "heathen" traditions of old Samoa were gradually transformed or eradicated. This gradual change meant that Mead still witnessed many precontact practices in the 1920s. Now, many struggle with the contradictions of trying to honor the remnants of old values from *fa'a Samoa* while simultaneously living with the realities of the new practices.

O'Meara (1990, pp. 69–70) provided a detailed account of how life has changed over the span of just one generation in a village he recently studied on the island of Savai'i. Included in these changes are: an increase in population,[2] less isolation, more travel (to Apia, American Samoa, and New Zealand), "new needs and desires," a "perception of scarcity" of material goods, more theft, and more family dispersion.[3] Many of these changes have led to a decreased sense of morale and an increased sense of frustration, particularly among the young who find certain aspects of village life less than satisfactory. As Meleisea and Meleisea (1980) argued, each "year thousands of young people graduate from secondary school but few find jobs and most feel that village farming is a poor return for their years in school" (p. 37).

In reference to how the "new" religion has transformed life in the village he studied on the island of Savai'i, O'Meara provided the following description of life in one of the more "fundamentalist" villages:

> Some church rules have a significant effect on village life. The Methodists are rather puritanical compared to most other denominations in Samoa. They allow precious little in the way of entertainment. The Church does not permit dancing at night (and discourages it during the day). It bans card playing of any kind, day or night, and strictly forbids the drinking of alcohol. . . . During the Christmas and New Year's holidays . . . several entrepreneurial families ran bingo games . . . by the light of gas lanterns in their homes or on the cricket pitch. The pastor banned the popular games after only two nights. (p. 44)

Admittedly, this village represents an extreme to which religious influence has transformed life in some villages, and the extent to which

---

[2]Meleisea and Meleisea (1980) noted that "the population of [Western Samoa] has grown from early 19th century estimates of 40,000 to approximately 160,00 in 1977, with something like a 3% annual increase. The consequence is a growing strain on land and inner reef marine resources" (pp. 36–37). From more recent records it is evident that the rate of population growth is declining. Between 1966 and 1971, the population increased by about 12%; between 1971 and 1976, the increase was only 4%; between 1976 and 1981, the increase was 3%; and between 1981 and 1986 it was only 1% (cf. Department of Statistics, Western Samoa, 1989). Although this "plateauing" should relieve some strain, it means that there is a population "bulge" now coming of age. In 1966, only 24% of the population were between the ages of 15 and 29 (31,510 people). As of 1986, 31% of the population were in this age range (48,495 people).

[3]O'Meara noted that a "generation ago families were still united, in good times and in bad. . . . Today migration is often expected and in many cases desired—yet it is also feared. Parents become separated from their children, friends are lost, and brothers and sisters scatter as they disappear into awaiting airplanes" (p. 70). He also noted that some of this family separation is caused by the necessity of family members going off to engage in wage labor to support their families. It is not uncommon for fathers, and especially sons, to work in Apia or American Samoa, but to only visit their families every few weeks.

Samoans no longer enjoy various forms of carefree activity and physical releases—such as dancing. Still, there are almost 24,000 practicing Methodists in Western Samoa (Meleisea, 1987a, p. 67). It is interesting to note in this context that Mead devoted a chapter to the importance of the *siva* (dance) "in the development of individuality and the compensation for repression of personality in other spheres of life" (p. 121).

Despite the prohibitions on hedonistic activities, some young people still engage in these emotional and physical outlets as part of a village "underground." O'Meara described some of this:

> All of these prohibitions apply to villagers wherever they are, but slackness sometimes creeps in when people travel to Salelologa, Apia, or overseas. Young men sometimes even hold clandestine poker games or drinking parties behind the village, though *matai* almost never attend these affairs. Villagers are generally willing to overlook breaches of proper conduct *as long as violators remain discreet*. If misconduct leads to a public outburst or to fighting, however, the village council imposes heavy fines on offenders. (pp. 44–45, italics added)

Many young people experience these restrictions as oppressive, particularly those who compare their own lives with the lives of young people in Western culture as portrayed in movies and magazines. For example, apparently many find the nightly curfew that has been instituted in most villages to be unreasonable (cf. Freeman, 1983, pp. 261–262). To violate the curfew brings a sometimes heavy fine for all except the *matai*, who are exempt because it is their duty to patrol the village (it begins at 9 or 10 p.m., depending on the village). O'Meara recorded that the curfew is in place "to assure that schoolchildren get to bed early; to prevent theft, *clandestine meetings of unmarried people*, and other elicit behaviors that might go undetected in the darkness . . ." (p. 56, italics added).

In relation to "Mead's Samoa," then, life in contemporary Samoa is obviously quite different—more repressive—as Freeman inadvertently brought to our attention. He also inadvertently brought to our attention the *deteriorating* situation facing young people there. As was seen in chapter 5, the *'aumaga* and *aualuma* now exist in name only, with few exceptions. Consequently, the primary village supports that once gave a sense of meaning and future purpose to the lives of the young have weakened. And, increasingly, the generations that once lived together in cooperation and mutual benefit now find family relationships increasingly strained. As Leacock (1987) argued, "the immediate source of stress for many young people is a perceived lack of empathy and concern on the part of stern and demanding parents, while parents who become angry over the perceived lack of proper respect on the part of the young may be excessively abusive'' (p. 182).

Hence, many of the young are in no-win situations. They are formally educated, and informally socialized, to expect certain material and emotional gratifications, but their prospects for realizing these expectations are often not good. There is now less opportunity for those who try to take the nontraditional, wage-labor route. For those who do take the traditional route, a less than satisfactory situation is increasingly encountered. In 1980, Meleisea and Meleisea characterized these problems in the following manner:

> The greatly increased educational facilities which have been available for Samoan children since the 1950s have created rising expectations among the younger generation, a dissatisfaction with village life and a longing for wage employment in Apia or a chance to emigrate . . . Returning migrants are another source of dissatisfaction with village life and the predominantly subsistence economy, as well as a source of new values and aspirations. (p. 37)

More recently, O'Meara (1990) summarized these problems by saying that "the cycle of *tautua* is broken" (p. 162). With this expression he meant that the traditional *quid pro quo* that existed in precontact society between the "young" and the "adult" has disintegrated, as he explained:

> The system of authority and service is organized largely on seniority—those who are younger serve and obey those who are older. Ironically, this kind of system is stable only as long as the participants are sure that it will continue. As long as the system is closed and the cycle secure, young people perceive their service as a tolerable burden, knowing that one day they will, in their turn, command the service of others.
>
> The cycle of *tautua* has been broken in the minds of many young people today. *New technology and a market economy make individual production, profit, and accumulation of wealth possible.* Many young people seek their futures in wage labour outside the village or even overseas, and they doubt that their own service will ever be repaid by a younger generation. In these circumstances, they no longer serve so gladly. (p. 162, italics added)

One attempt to augment the sense of meaning and identity among young males was to increase the number of *matai* titles granted. According to O'Meara (1990), the "number of *matai* has been growing rapidly over the last three decades, so that now most adult men hold titles. In rural areas such as Savai'i, 75 percent of all men 21 years and older are matai.⁴ That is more than double the percentage of thirty years ago" (p. 151). But apparently this "oversupply" has had the effect of diminishing the importance of, and respect for, these titles. For example, many young men have been given titles without the traditional "apprenticeship" and therefore lack many of the skills necessary to exercise their power wisely. Evidently, some of these less qualified *matai* are involved in more conflicts with both elder *matai* and younger untitled men, in part because some of these young *matai* apparently do not conform to the consensus-building conventions of *fa'a Samoa* (cf. Leacock, 1987).

So, where does this leave the young person who is coming of age and looking to develop a sense of identity as an adult?

Freeman would have us believe that young people in Samoa have always had problems with their adolescence because of the nature of traditional Samoan society, but many observers disagree with him (Leacock, 1987). In her analysis of the problems facing young people in contemporary Samoa, Leacock argued that Freeman has been remiss in his "failure to deal seriously with the complexities of recent socio-historical change" (p. 177). In addition, she contended that "it is a serious misrepresentation of the situation to say that youth difficulties simply follow from the authoritarianism of the Samoan past. *Instead they follow from the fact that the nature of youth as a life period has been transformed in many ways*" (p. 186, italics added). Similarly, in response to Freeman's citation of suicide statistics, Shankman (1983) argued that a "more plausible explanation might lie in the modernization of Western Samoa and its interaction with traditional Samoan culture, since there are different rates of suicide for Western Samoa and American Samoa, where the suicide rate is considerably lower" (p. 52).

The issue of suicide and its recent rise in Western Samoa strikes at the heart of the dilemma facing those now coming of age in Samoa. Moreover, when we examine the nature of the suicide problem, we see more clearly a principal flaw in Freeman's critique of Mead's coming-of-age thesis.

The most comprehensive analyses of suicide in Western Samoa were conducted by Bowles (1985) and MacPherson and MacPherson (1985). According to Bowles, the suicide rate was low in the early 1960s (about 2 per 100,000) and it rose slowly until the mid-1970s (about 10 per 100,000). After this, it rose dramatically in the early 1980s and began a decline again by the mid-1980s. Averaged between 1981 and 1983, Bowles reported a rate of 22.6 for the entire population, with the following breakdown based on age and gender: For males, the total rate was 31.3, but for males 15 to 24 years of age it was 71.0 and for males 25 to 34 it was 75.6; for females the total rate was 13.3, but for females 15 to 24 years

---

⁴According to O'Meara (1990) "less than 0.02% of all *matai* are women" (p. 33).

of age it was 35.8 and for females 25 to 34, it was 20.4 (note that the actual total number suicides is 106 for 3 years). More recent figures reveal that in spite of a prevention program and an apparent decline, the problem persists. In 1990, there were 27 confirmed suicides, constituting a rate of 16.9 for a population of 160,000 (Aiavao, 1991).

Based on their sociohistorical analysis of Samoan culture, MacPherson and MacPherson (1985) concluded that although forms of suicide have probably existed throughout the history of Samoan culture, "suicide . . . does not seem to have been an institutionalized response" (p. 37) to problems. They also concluded that the recent suicides constitute forms of "anomic" suicide. In a similar vein to the analyses presented here, MacPherson and MacPherson developed a "blocked opportunity model" that argues that suicides since the mid-1970s can be explained in terms of rising expectations regarding affluence and personal freedom produced by mass education, the media, the wage economy, and emigration prospects. These expectations have been unmet for large numbers of youth who face limited opportunity for paid employment, and who in the mid-1970s had their emigration possibilities severely limited. But, instilled psychologically with alternative conceptions of what life could be like, many are now forced to live in villages with limited economic prospects as well as with families demanding a blind obedience to traditional customs. Accordingly, many of the cases that MacPherson and MacPherson examined involved a thwarted attempt by a young person to exercise some type of autonomy disapproved of by a parent or a *matai*. Hence, many of the suicides and attempted suicides appear to constitute a "statement" by a frustrated individual against what is perceived as an intransigent traditional authority figure. Evidently, much of this desire to exert independence has been stimulated by exposure to Western influence, especially education and the mass media. It is of interest to note Mead's predictions regarding the problems that would emerge in Samoa when individual choice was demanded by the young (e.g., 1928, pp. 169–171).

O'Meara provided a similar explanation for the recent rise in suicide:

> The correlation of the suicide epidemic with the rapid modernization of the post-independence era in Western Samoa is no accident. The desires and expectations of young people are changing very fast. Radios, movies, television, videos, and New Zealand-style education all give people new ideas and new dreams. . . . The closeness of that other world sometimes makes patience difficult, yet the actual pace of social and economic change makes the fulfilment of these expectations very unlikely. Rapid change alone is not the problem. Suicide is uncommon in American Samoa, where change has been far more rapid and dramatic than in Western Samoa, and where half the population is made up of migrants from Western Samoa. The real

problem appears to be *uneven* change, specifically the lag between young people's growing expectations and the social and economic realities in which they live. (p. 111)

Given the no-win situation that many young Samoans face, in a sense it is surprising that there have not been more suicides. It is also estimated, however, that about twice as many *unsuccessful* suicide attempts take place (Bowles, 1985, p. 17). More recent reports suggest that the frustration among the young is being manifested in other ways. For example, marijuana smoking has become "a problem in Western Samoan schools, especially in rural areas . . . Some schools [report that] drug abuse has replaced cigarette smoking and beer drinking as the main offence leading to expulsion" ("Drug Abuse Hits Schools," 1991). Foreboding a more serious problem, it has been reported that "Western Samoa's Police Commissioner . . . has called for the establishment of a special jail for young people" ("Special Youth Jail?," 1992).

We can put the problem of suicide in Samoa into perspective in two other ways.

First, it appears to have the qualities of an epidemic, which means that potentially there is a "cure." The peak of this epidemic in the early 1980s appeared to represent a "contagion" of collective behavior from which it became "fashionable" to drink the weedkiller "paraquat."[5] The death from this poison is agonizing and prolonged (sometimes more than a week), thereby ensuring that the authority figures "responsible" are made to suffer also. Of the 49 known suicides in 1981, 39 were from paraquat poisoning; in 1990, 19 of the 27 suicides were from paraquat (Aiavao, 1991). It is believed that many of the attempts are not genuine suicide attempts, but simply either a means of challenging authority or a passive–aggressive attempt to hurt someone. Apparently, some suicide victims take only a small amount, thinking that they will just get sick, but even a small amount of paraquat can result in death. Access to paraquat has still not been restricted, despite 277 known deaths from it since 1973 (Aiavao, 1991).

Second, we can compare Samoan suicide rates with those from other countries, and take lessons from this comparison. It is often said that Western Samoa has the highest rate in the world; in fact, even some guidebooks now make reference to it (e.g., Swaney, 1990). But, Samoa should not be singled out so readily. In point of fact, the suicide rate for young people increased in many countries around the world between the 1960s

---

[5]There is considerable evidence from U.S. studies that when a suicide is publicized by the media, the rate of suicide increases among people with similar characteristics to the person who committed suicide (Phillips, 1979; Stack, 1987).

and the 1980s, including major industrialized nations. In Canada, for example, it rose from 5.3 per 100,000 for 15- to 19-year-olds in 1960 to 20.2 in 1986, and from 12.3 for 20- to 24-year-olds in 1960 to 32.8 in 1986 (Beneteau, 1988). Moreover, if we look at Canada's indigenous population, we find a rate of 100 per 100,000 among Native Canadian men ages 15 to 19. This rate is stable and is clearly above the rate for Samoan men of the same age. No doubt, similarly high rates can be found among groups that have been disenfranchised in some way. But the rates should also be particularly high among those who have been culturally disenfranchised and who are attempting to come of age, as is the case with both Samoan and Native Canadian young people (cf. Rubinstein, 1985, on suicide in Micronesia).

From these two perspectives we can see that Samoa is not alone with its suicide problem, and that the remedy to this epidemic somehow lies with a lack of a "franchise." To effect a "cure," the questions that must be answered pertain to the exact nature of the relationship between suicide and the lack of a franchise, including how to "re-enfranchise" the young, given that without a franchise the young have no representatives of their own to speak for them on policy matters. I return to this issue later in this chapter.

These suicide statistics likely represent the "tip of the iceberg" with respect to the lack of personal meaning and sense of alienation experienced by many of the young. The alienation associated with this lack of meaning has been expressed by a number of Samoa's poets (Malifa, 1975; Petaia, 1980; A. Wendt, 1974). A. Wendt's (1973, 1977, 1979) novels capture some of this alienation and despair, as does a recent movie based on one of his novels (*Flying Fox and the Freedom Tree*; see Robie, 1991, for a review). The poem "Kidnapped" especially captures some of the feelings produced by an "alien" and alienating educational system (Petaia, cited in A. Wendt, 1974):

> I was six when
> Mama was careless
> she sent me to school
> alone
> five days a week
>
> One day I was
> kidnapped by a band
> of Western philosophers
> armed with glossy-pictured
> textbooks and
> registered reputations
> "Holder of B.A.
> and M.A. degrees"

I was held
in a classroom
guarded by Churchill and Garibaldi
pinned-up on one wall
and
Hitler and Mao dictating
from the other
Guevara pointed a revolution
at my brains
from his "Guerilla Warfare"

Each three month term
they sent threats to
my Mama and Papa

Mama and Papa loved
their son and
paid ransom fees
each time

Each time
Mama and Papa grew
poorer and poorer
and my kidnappers grew
richer and richer
I grew whiter and
whiter

On my release
fifteen years after
I was handed
(among loud applause
from fellow victims)
a piece of paper
to decorate my walls
certifying my release

As just mentioned, one way in which many young people once "escaped" the frustrations of living in contemporary Western Samoa was to emigrate. Emigration from Western Samoa has been primarily to New Zealand, its colonial "ruler" since 1914. The Samoan community in New Zealand now numbers some 100,000 (Meleisea, 1987a, p. 161). For some time, the young could freely move to New Zealand to work temporarily, but in the mid-1970s the New Zealand government became concerned about "overstayers," and have since restricted work permits. Since then, immigration has been restricted to those who "have a guarantee of a job in New Zealand," or who qualify under a special quota system that allows 1,100 Western Samoans per year on family or humanitarian grounds

(Immigration Division, New Zealand, 1982). The resulting work scheme has led to a trickle of persons being admitted for temporary employment: 11 permits in the first year of the program (1977), none in 1983, and 3 in 1984 (Labour Department, Western Samoa, 1984). According to MacPherson and MacPherson (1985), this move by New Zealand has constituted a serious blow for many young people in terms of their plans and expectations, particularly because this period in New Zealand had come to constitute something of what E. Erikson (1968) called an "institutionalized moratorium," in this case a *Wanderschaft*, in much the same fashion that traveling to Europe constitutes one for young North Americans (cf. Norton, 1984).

This chapter began with the assertion that young people have been largely forgotten in the furor among social scientists over the Mead–Freeman controversy. But, of the limited attention given young people, young women have been given even shorter shrift. Indeed, there is an irony in the fact that Mead's study of a few female adolescents in the 1920s has led to so much attention being paid to the reputations of social scientists and so little attention being paid to young women struggling to come of age under the conditions just described. Moreover, in relation to young men, young women appear to have lost even more.

As was seen in chapters 4 and 5, young women have lost the organization—the *aualuma*—that, in precontact Samoan culture, gave them solidarity and semi-autonomous sexuality. In what appears to have been an attempt by missionaries to establish a patriarchal nuclear family structure with its associated morality, the role of women in Samoa was targeted for dramatic change. By the time of Mead's study, the chosen few who maintained their chastity were trained to be pastors' wives in the boarding schools, whereas the remainder were left mainly on their own, forced to face an adult world that evidently was increasingly hostile to expressions of their autonomy and individuality. According to O'Meara (1990):

> contemporary Samoans apply a familiar 'double standard' toward unmarried males and females. Young, unmarried men are allowed (or even encouraged by each other) to have affairs (though only with unmarried women). The same activity is forbidden of girls, whose virginity and reputations are social assets as well as moral virtues. (pp. 107–108)

Young women also seem to have lost much of the moratorium period that Mead described, where they could enjoy themselves with a "time out" before taking on adult responsibilities and family commitments. O'Meara (1990) wrote that in the religiously conservative village he studied "young women are . . . closely guarded by other family members"

(p. 104) and that "women's daily activities are delineated partly with an eye on keeping [them] close to home" (p. 101). He also noted that most "families go to great lengths to guard and restrain their young girls. In the face of such constant chaperoning, most girls have to actively conspire in order to meet privately with a lover or a suitor" (p. 108). However, he also noted instances of premarital affairs and births, and he speculated that much of this constitutes a form of rebellion against parental authority and an attempt to hurt parents in a passive–aggressive manner, similar to the suicide attempts discussed earlier. His sources in that village told him that "sexual escapades of this kind are an increasing problem" (pp. 108–109).

O'Meara also gave us a glimpse of what daily life is now like for women living in Western Samoan villages in terms of the division of labor that has evolved:

> Males of all ages find more time to play or rest than females. In the late afternoons the young men and boys can usually be found relaxing or playing cricket on the village green, playing rugby on a sand flat exposed by the tide near the lagoon shore, or playing volley ball at the school grounds. A few young women join in the volleyball, but most remain at home, cooking and tending to the many children or countless household chores. Even into advanced middle age, women are expected to keep busy (though not all do). Meanwhile, their husbands often sleep, play dominoes, or discuss the intricacies of the latest social or political manoeuvre, leaving their own household duties half-completed. Women sometimes express resentment towards men for this unequal division of labor. (p. 70)

We can see evidence that as power shifted in social relations over this century, at the microlevel of domestic roles, men appear to have gained at the expense of women. Furthermore, at the more macrolevel of political participation and power, women appear to have gained little from recent cultural change. Again, O'Meara gave us a glimpse of this:

> Like men, women gain more control over their lives as they grow older. Even if a woman outlives her parents, however, she still may not fully control her own life until she is an elderly widow. There are exceptions, of course, when a woman rises to authority in her own household, her extended family, or even in village affairs. I know of no such cases in Vaega, however [the village he studied]. A very few women ever rise to national political prominence, but even though these women are exceptionally talented, they usually gain their initial opportunities through genealogical or marital relationships with powerful men. (p. 70)

With few women in positions of formal power to speak for them, it is unlikely that the rights—the franchise—of young women coming of age will improve. Thus, we can identify a "crisis" in gender roles, whereby the gulf or disparity between the status of women and men in Samoa has widened, with women experiencing a sharp decline in status. This crisis seems to be contributing to the growing social malaise there, particularly among young women.

DEALING WITH AN UNFOLDING TRAGEDY FACING WESTERN SAMOAN YOUTH

Leacock (1987) is one of the few social scientists to have focused on the problems facing contemporary young Samoans. She concluded that "all indications are that adolescence has indeed changed from being a period of relatively little stress . . . to being highly charged with stress manifested by delinquency and suicide" (p. 181). Freeman somehow missed this in his critique of Mead's work. But, as has been seen throughout this book, his critique is flawed because he misrepresented crucial elements of both her position and Samoan cultural history. Inevitably, then, the "mission" he set for himself will fail.

In spite of these misspent efforts, there is at least one positive consequence of the worldwide attention that Freeman has drawn to Samoa; namely, awareness of the desperate situation facing many young Western Samoans attempting to come of age in a culture that has been inundated with Western influences. In the first part of this chapter, we examined this serious situation. In this last part of the chapter, I make recommendations regarding what might be done to alleviate it.

As is the case in many Third World nations, young people have been culturally disenfranchised because of the "imperialism" of Western religious and economic practices. In this light, perhaps the best that may come of the Mead–Freeman controversy is an awareness of the trends that have robbed young Samoans of much of their cultural heritage. Then, perchance, the attention paid to Freeman (and his attack on Mead and American anthropology) can be diverted to finding solutions for the very real problems facing those now "coming of age in Samoa."

Policy Recommendations

Leacock (1987) noted that the problems facing young Samoans are shared by young people in many countries where "traditional cultures" have been seriously distorted by the imposition of international capitalism: "Elsewhere in the world, delinquency, suicide, and other escalating problems of Third World youth are seen as arising from new social and economic conflicts, and from the malaise and hopelessness associated with such difficulties as school failure, unemployment, and loss of cultural identity" (p. 178).

Simply stated, the introduction of capitalism into "traditional cultures" often culturally disenfranchises the young because it breeds individualism and selfishness. With a decline in community orientation among adults, the young are left with less guidance, and fewer of the fruits of collective labor are shared with them.

In order to avoid the problems engendered by Western economic and cultural imperialism, ideas for the solutions to many of the problems discussed here should come from within Western Samoa, so it is perhaps presumptuous for a Westerner such as myself to "interfere" in this respect. On the other hand, with due respect I offer some of my thoughts on this matter, accumulated as I investigated the Mead–Freeman controversy. Accordingly, I discuss what I think might be done in rectifying the transformations that have taken place in Western Samoa over the past 150 years in the areas of religion, education, and the economy.

Religion

The missionaries were obviously well intentioned, but it is clear from what we saw previously that their influence is partly responsible for the situation now confronting many young Samoans. In this light, the responsibility falls to some extent upon the clergy now in Samoa to address the cultural disenfranchisement of young Samoans in whatever ways they can. For example, given their integration deep into the life of most villages in Samoa, it can be argued that the onus is on pastors and religious authorities to restore the dignity previously accorded to coming of age, and to re-institutionalize structures such as the *'aumaga* and *aualuma*. Energy and resources could be directed toward providing more guidance and structure to those coming of age, thereby enhancing the sense of meaning and purpose in life of young people. These leaders should take heed of their reputation, as described by Crocombe (1989): "Christians have not necessarily practised all they preached, and allegations of public exploitation by ultraconservative churches in Samoa . . . contain elements of truth. The average stomach girth of religious ministers . . . is sufficient indication" (p. 74).

My recommendation is for contemporary religious leaders in Samoa, who owe their callings to the efforts of missionaries with a "vision," to embark upon a new "mission" with a new "vision" —to save the young from the cultural disenfranchisement engendered by the international economy. In view of the evidence presented in this book, the "new missionaries" could set out to re-institutionalize benign and meaningful passages to adulthood, replacing the institutions ravaged by the "old mission."

Education

Most contemporary institutions in Western Samoa are based on the British model, by way of New Zealand's influence. This is certainly the case with the educational system,[6] which has been criticized because of its irrelevance to Samoan culture. Writing in the 1970s, Pitt and MacPherson (1974) said the following about that system:

> While Western Samoa was a Trust Territory administered by New Zealand, the philosophy behind the education system did not promote maximum teaching efficiency. It was primarily concerned with giving Samoans a European-style, basically humanistic education. The Samoan context was not taken into account nearly enough. For example, children used cultural materials based on meaningless foreign examples—like the readers which told of robins in the snow or train trips to London. Undue emphasis was placed on the use of English which even the teachers did not fully understand. (pp. 99–100)

Efforts have been made recently to correct this inadequacy. For instance, with respect to relevance of curriculum, currently under way is a "Samoanization of assessment procedures and more comprehensive curricula at Junior Secondary and Senior Secondary Levels" (Department of Economic Development, Western Samoa, 1987, p. 173). One of the five explicitly stated objectives and goals of educational policy is "to prepare Samoans particularly the youth to handle and appreciate the challenges and pressures of an increasingly modern society" (Department of Economic Development, Western Samoa, 1987, p. 180). At the primary level, an objective is "to cater to [students'] personal and moral development so that they are induced to think rather than by rote as previously done," while at the secondary level an objective is to facilitate the pursual of "a 'student's own' respective area of choice" (p. 181).

More recently, a system has been adopted whereby students are graded on a 9-point scale in each subject. This system eliminates the conception of pass–fail, and introduces the notions of satisfactory–unsatisfactory ("Education Director," 1991). The intention is to stop the practice of label-

---

[6]Education in Western Samoa is not compulsory. As of 1981, 81.8% of the 5 to 19 age group were in school, up from 72.1% 10 years earlier. In terms of an age breakdown, 81.1% of 5- to 9-year-olds were in school in 1981, as were 97.5% 10- to 14-year-olds and 64.9% of 15- to 19-year-olds. The largest increase in enrollment by age group was the 15- to 19-year-old group, 43.8% of whom were in school in 1971. In addition, many Samoans have been educated in New Zealand, particularly at the secondary and tertiary levels.

ling a student as a failure. The label of "failure" has serious consequences in any culture, but it is especially deleterious in a shame-oriented culture.

Several projects have been launched to deal with the problem of delivering a useful education to young Samoans. One such project, the "Intensification of Vocational and Skills Training Programme," has as its priority out-of-school youth who have high employment and skills potential. It also has the mandate to conduct market studies to assess the employability of youth. These apprenticeship plans, however, have been hampered by a lack of funds and a lack of skilled personnel who are willing to manage the program for the wages offered or to teach for the meagre remuneration the government can afford. One program had 83 apprentices in 1983 but only 60 by 1987 (Department of Economic Development, Western Samoa, 1987, p. 178); another, created by the Apprentice Act of 1972, had only two positions in 1984 (Labour Department, Western Samoa, 1984).

Thus, inadequate finances set up a "loop" whereby a problem feeds itself.[7] In fact, in 1985 the entire education budget was only 7.7665 million Western Samoan *tala* (about $4 million U.S.) and this constituted about 25% of the government expenditures for that year (Department of Economic Development, Western Samoa, 1987, p. 229). In Western societies, one school alone might operate on such a budget! Besides mission schools, the only major form of relief comes from the Peace Corps volunteers who make up a proportion of the 1,500 teachers who teach some 50,000 students (Douglas & Douglas, 1986, p. 510).

In 1984, the National University of Samoa was founded, but it has experienced difficulties getting on track as a degree-granting institution. As of 1987, it still provided only a University Preparatory Year to prepare students for undergraduate studies overseas, mainly in New Zealand, Australia, and Fiji (Department of Economic Development, Western Samoa, 1987, p. 129). By 1987, 201 students had graduated, with 126 moving on to university programs. Plans are in place to establish degree programs in education, commerce, accounting, arts, and science.[8] The University

---

[7]The bottom line of the problem facing the government in attempting to implement many policies is, of course, a lack of finances. For example, the Ministry of Youth, Sport, and Culture must function with little money beyond its own operating budget. Recently, two new programs were added, but they had meager budgets: the Sports and Cultural Development Programme for Youth was to receive 11,000 *tala* per year (about $5,000 U.S.), whereas the Youth Training program was to receive 77,000 *tala* in the first year and 27,000 in the second year (Department of Economic Development, Western Samoa, 1987, p. 195).

[8]If the National University of Samoa could expand to provide undergraduate and graduate degrees, Western Samoa would perhaps become a more vibrant intellectual and cultural centre. This could be accomplished with the assistance of some the hundreds of thousands of academics around the world, in the form of sponsorships, exchanges and visiting professorships. Consequently, the National University of Samoa could not only educate many its own citizens for the more highly skilled business, professional, and administrative positions, but it could address its serious brain drain problem.

of the South Pacific also has a small campus near Apia that provides courses in agriculture.

Although the official objectives of the educational policy cited here are reasonable, several problems are evident. First, although it is difficult to be too critical of efforts to raise the level of educational attainment of a nation, mass education should not be viewed as a panacea for dealing with social and economic problems. The model for this educational system comes from Western societies, which themselves have enormous problems in terms of their own educational systems (e.g., Collins, 1979). Wanting to "blindly" educate everyone through secondary school (and university) may be well intentioned, but the side effects of such social engineering must be recognized and addressed. Again, this is not to say that education itself should be discouraged; rather, education without clearly articulated and integrated goals has not been an effective mechanism for economic development anywhere in the world.

To reiterate, although mass, Western-style education undoubtedly has certain benefits, it may also have some deleterious effects. These negative side effects include an alienation associated with the creation of expectations that cannot be met, and the cultural disruption that results when values are inculcated that conflict with the circumstance students face in their home life and when they leave the school. Certainly, the intention to create a heightened sense of awareness and cognitive functioning is a laudable one. But, if the cultural setting is not appropriate for the forms of cognitive and personal development stimulated by Western-style, mass educational experiences, policymakers must be certain that they also inculcate the psychological qualities that enable individuals to deal and live with these conflicts. From the point of view of social engineering, there is a serious problem with the lack of knowledge regarding how to systematically inculcate the psychological qualities that "immunize" people from the alienation often created by mass education (cf. Côté & Levine, 1987).

Human capital theory, developed by economists, advances the view that "individuals are like a piece of machinery, a capital good, and can increase their value in the labor market by increasing their education, especially training in occupational skills" (Ballantine, 1989, p. 301). Although human capital theory has informed the educational policy behind the expansion of educational systems in industrialized countries since the 1950s, it has been exposed as being of limited validity. This is especially the case beyond the point of achieving functional literacy among the population, so that basic commerce takes place, and beyond producing a *small number* of highly trained workers, so that specialized functions are performed (cf. Lockhart, 1971, 1975; Porter, 1984). Although a literate population is beneficial to an economy, much of the learning that takes place beyond the point of basic literacy is not applicable to most jobs in the industrial and service sectors of the labor force (Berg, 1970; Collins, 1979). Instead, what results is an inflation of educational credentials—credentialism—that simply sets expensive and arbitrary standards for access to jobs. Students are often merely kept "busy" for a requisite number of years, and "warehoused" in schools until they are certified as "properly educated."

Policymakers must deal with the fact that modern Western education—which emphasizes individualism, careerism, and consumerism—does not correspond with the values of *fa'a Samoa*. Moreover, there is no absolute reason for Samoa to participate fully in the international economy; in fact, by doing so it appears that much will be lost and given away, as has been the history of Samoa's contact with the *palagi*. The contradictions imposed upon Samoan youth are illustrated in the following statement of official policy, from which some programs discussed earlier were fashioned:

> Such programmes shall aim to develop and transform the youth into creative, productive, and socially responsible citizens and prepare them for the task of nation building. In recent years Samoans want wider opportunities for self-expression and individual achievement. Youth counselling and guidance programmes shall be geared toward value orientation and moral transformation and address such issues and problems on adolescence and reluctance to accept traditional roles by a growing number of young Samoans. (Department of Economic Development, Western Samoa, 1987, p. 191)

Within this policy statement are the contradictory goals of instilling individualism while coaxing young people to accept traditional community roles. Although the need for such programs exists, their goals must be carefully considered. For example, to deal with the need of combining Western values with Samoan traditions, thought might be given to establishing new customs, such as a synthesis of the *'aumaga* and *aualuma* with secondary education. Accordingly, young people primarily would learn the values and skills of their ancestral culture, but secondarily they could develop non-Samoan knowledge. This would potentially revitalize the village economy and make wage labor less necessary, as these youth groups return to carrying out collective projects. In addition, the need to leave the village in search of wage labor could be reduced if these new youth structures taught craft skills and fed local self-sustaining industries. The consequences of re-mobilizing youth labor include reinstating local village autonomy, restoring the dignity to being responsible for one's community, and returning a sense of meaning to those coming of age. The Western Samoan government can contribute to this by undertaking more initiatives like the "Rural Development Programme." This program has benefited women (cf. Muse, 1991; Natarajan, 1983), and similar programs can be directed to explicitly meet the needs of young people.

This is not to say that Samoans should abandon the world educational/economic system, or that the world educational/economic system should abandon Samoans. Rather, given the uniqueness and potential of the Samoan situation, one could hope that new models could be developed to help Samoa and other developing countries avoid the traps that the educational/economic morass creates. Salmi (1992) discussed strategies for dealing with such problems as they apply to higher education.

The Economy

As noted previously, around the world it is generally assumed that formal education constitutes a period of preparation for the labor force. On the basis of human capital theory, it is further assumed that an "investment" in education by the state and the individual will lead to some tangible return for both (cf. Pitt, 1970). As also noted previously, these expectations have been found to be unrealistic as the solution to economic problems, particularly in an "undeveloped" country like Western Samoa. It is openly admitted by the Western Samoan government that "at present, there is no coordinated framework for national manpower planning. . . . Overall the training of personnel for manpower requirements for the public and private sector remains an ad-hoc exercise" (Department of Economic Development, Western Samoa, 1987, p. 22). This statement characterizes the "patchwork" economic system in Western Samoa. Nearly all observers agree that there are significant problems with planning and coordinating efforts in several sectors of the Western Samoan economy. By the same token, this is an area in which local solutions can be attempted.

The employment that is available, is generally of low wage. The minimum wage was 55 *sene* per hour (about 25¢ U.S.) in 1984, and the average salary for those in paid employment was about 2,300 *tala* per annum (Labour Department, Western Samoa, 1984). There appears not to be much variation in salaries, primarily because there is not much variation in the type of employment available—it is largely clerical and service, with some light manufacturing (cf. Meleisea & Meleisea, 1980; "Western Samoa Push For Investment," 1992). Of the 21,400 known to be in paid employment in 1983, one third were public servants (Labour Department, Western Samoa, 1984). Moreover, there is some question regarding the appropriateness of the training of public servants:

DEALING WITH AN UNFOLDING TRAGEDY 143

> Western Samoa's labour market is characterized by some serious imbalances. First and foremost, while there is a growing number of General Arts & Science graduates employed in the service, there is a lack of specialists particularly in the scientific field (Department of Economic Development, Western Samoa, 1987, p. 22)

As for unemployment, it is difficult to get accurate statistics because there are no unemployment insurance or welfare systems. Leacock (1987) pointed out that youth unemployment has "no counterpart in societies based on subsistence farming, where the economic future of young people [is] assured." But around the world, in countries forced into the international economy, "youth unemployment is on the increase, and everywhere with deleterious psychological effects" (p. 178). The Western Samoa government does not keep track of unemployment statistics, partially for the following reasons:

> Employment here is not gauged in terms of formal labour requisites. And the figures do not denote full time employment. In fact, there is considerable part-time employment. Employment in the Western Samoan sense, is employment of virtually any sort, from working as a teller in a commercial bank to growing bananas in the backyard. (Treasury Department, Western Samoa, 1988, p. 10)

This report, compiled to attract foreign investment to Western Samoa, goes on, however, to admit that "in other words, there is gross underutilization of the country's resources, especially where labour is concerned" (p. 10). But it goes on to argue that the massive "underemployment" is an asset because of the labor force awaiting the development of capital enterprises. Apparently, however, these efforts to entice multinationals are meeting with some success (e.g., McCabe, 1992). On the other hand, multinationals will only be attracted by very low wages and will offer jobs requiring low levels of skill. For example, the automotive manufacturer Yazaki recently opened a branch plant in Apia, employing some 1,500 people. This operation assembles electrical components. The managing director of the Yazaki Australia, who set up the project, is quoted as saying the following about Samoan workers:

> We have recruited, selected, inducted, trained and now promoted many people, and find them to be excellent. . . . In some sense, our assembly operations are akin to handcrafting. It therefore follows that a nation with a heritage of basket-weaving and thatching will be very adept. Attributes such as dexterity, suppleness of limbs, basic body strength, good eyesight, superior binocular vision, superior acuity, accurate colour recognition, good co-ordination and motor actions are important. (McCabe, 1992, p. 39)

Note that there is nothing in these attributes that reflect the need for a secondary education. Note further the "psychosocial attributes" that are identified as making Samoans "good workers":

> He said Samoans had a pride in achievement, accepted authority, were aware of the need to safely preserve materials, produce error-free outcomes, report suspected faults, practice error containment and self-inspection functions, and to cooperate through "buddy checking" systems—for all this was readily facilitated by people "whose racial upbringing is based on harmonisation." (McCabe, 1992, p. 39)

In spite of what has to be a patronizing relationship with these multinationals, the government of Western Samoa faces a very serious economic problem as a member of the international community. According to Labour Department statistics, in 1981, 87,110 citizens were over the legal working age of 15, constituting 55.8% of the population. Of these, 41,506 were "economically active" (85% of the "economically active" were men). As the earlier statistics indicate, however, only about one half these would be in "paid employment," the rest presumably working in their *aiga* plantations and related activities (Labour Department, Western Samoa, 1984). These figures presumably include those driving the estimated 500 taxis in Apia ("Taxi License," 1991), a city of only 45,000 inhabitants (O'Meara, 1990). In any event, putting these figures together means that only about 25% of the adult population are in the wage economy labor force.

As for youth unemployment, if the general level of "unemployment" is about 75%, we can be sure that it is significantly higher for the young because this is consistently the case in other countries. In terms of dealing with the problem of youth unemployment, the government appears to be counting on the impact of foreign investment: "obviously we have to further develop industry and service sectors which will eventually utilize the talents of our youth" (Treasury Department, Western Samoa, 1988, p. 10). Unfortunately, as we have just seen, most foreign investment will be exploitative. A recent wildcat strike at the Yazaki plant signals that the promise of economic salvation through cooperation with multinationals is probably illusory (North, 1993).

Granted, the young can work at "unpaid labor" for their *aiga*, but as we have seen, it is questionable how satisfactory this is for the increasing number of high school graduates. Even many of those who do not aspire to work in the wage economy, and instead work within the traditional *aiga* system, seem to develop an appetite for consumer goods—especially "identity" items like running shoes, ghetto blasters, and the latest Western fashions—that they view in the media or see on trips to

Apia or American Samoa. Obviously, however, unemployment and low wages provide little cash with which to buy such goods. Concern has been expressed about this problem from many sources, and locally there is an impatience with the inaction in dealing with it, as this recent newspaper commentary indicates:

> while so many people are thinking about getting a job and earning money for the family one cannot help but wonder just what the rest, or should one say the huge majority, of school leavers without jobs are going to be doing. It looks like there will be a whole lot more young people around here with nothing to do and the increase in juvenile crime is about to take off. . . . Such a pity that with all the talk we keep hearing of how we look after our own no one really wants to look at this problem any longer. . . . ("Column Seven," 1991)

Clearly, it is wrong to say that the government, or Samoans in general, do not care about what is happening to their young. Official government policy is informed by *fa'a Samoa*, which includes the principle of care for everyone. As we have seen throughout this book, however, the problem is that the entire society is being swept by the currents of change, so it is difficult to focus attention on what to do about the problems affecting those coming of age. MacPherson (1988) provided a detailed analysis of the economic problems in Western Samoa brought on by this change and how these problems can be dealt with by harnessing traditions, rather than rejecting them.

THE CULTURAL HERITAGE OF THE YOUNG: INDEPENDENCE OR DEPENDENCE?

Western Samoans are proud of their political independence and the fact that they were the first nation of the South Pacific islands to break its colonial dependence. Ironically, it may be economics, not politics, that returns Western Samoa to a colonylike dependence. This may happen if multinational corporations are allowed to exploit Samoa's "cheap labor." We can hope that Western Samoa may also be the first to resist this form of dependency. To do so, however, will require explicit policies designed to instill local autonomy.

As argued previously, it appears that the assumptions driving many current policies in Western Samoa are based on "human capital theory" and "modernization theory" (cf. Pitt, 1970). Human capital theory predicts that economic activity is stimulated in direct proportion to the amount of education obtained by workers in the labor force—that "investments" in education yield dividends for the economy. Similarly, modernization theory predicts that economic activity is stimulated when certain "beliefs, values, and behaviors" are adopted by a population; namely, "diligence, rational calculation, orderliness, frugality, punctuality, and achievement motivation—and new social values such as meritocracy, getting ahead on one's own ability" (Ballantine, 1989, p. 301).

Although both theories appear reasonable at a common sense level, both have been questioned because of their limited applicability. For example, meritocratic principles remain an ideal rather than a reality in most advanced industrial nations, especially capitalist ones (Krauze & Slomczynski, 1985). In addition, investments made in education in many Third World countries are often lost through "brain drain," as those with higher levels of education leave the country because of low wages and unemployment (Ballantine, 1989). Western Samoa currently has a serious brain drain problem (e.g., Crocombe, 1989). Finally, knowledge on its own does not stimulate economic activity. Rather, only specific forms of knowledge that are directly relevant to fiscal realities stimulate economic activities. For the majority of jobs in most countries, this level of knowledge is reached with a grade nine education and functional literacy (cf. Porter, 1984). Being well read and worldly—having Western-style "cultural capital"—may be personally gratifying, but on their own these attributes do little in terms of the day-to-day realities in the wage labor force, especially in countries like Western Samoa where most jobs involve service, clerical, or manual labor.

Samoans were promised 150 years ago that if they embraced Christianity, they would inherit the material comforts of the *palagi*. They were assured in this century that if they embraced capitalism they would inherit the affluence of *palagi* societies like New Zealand and Australia. Like most other countries touched by Western values, Samoa has been caught up in the "myth of progress," based on the assumption that "growth is good for society and the individual" (Ballantine, 1989, p. 308). Unfortunately, in believing the myth, countries like Western Samoa simply become "small fish" in a big "economic pond." They are useful to the "big fish" in that pond, but only as consumers, sources of cheap labor, and a source of natural resources. On the other hand, in a sense Western Samoa has been fortunate because it does not have many natural resources to attract the attention of these "big fish" (except its hardwood lumber, but most of the pristine stands have already been cut; see Cox & Elmqvist, 1991). Consequently, it has been left largely alone. If it were resource rich, I doubt if it would now be an independent nation—New Zealand would not likely have so easily consented to its independence.

A perspective that sheds light on this discussion is "dependency theory" (see Allahar, 1989, for a review). This theory takes a global perspective and examines the history of contact among nations. When viewed from this perspective, investments in mass education made with the intention of fostering "modern values" pale in comparison to the role played by such things as "a nation's structural position in the world economy, trade flow, dependence on primary product exports, state strength, degree of foreign investment, and the presence of multinational corporations" (Benavot, cited in Ballantine, 1989, pp. 301–302). Moreover, the role played by education in economically dependent countries is not so much the direct stimulation of the local economy as it is the inculcation of attitudes that nurture an acceptance of the world economic order, particularly among the political elite (cf. Crocombe, 1989). As Ballantine (1989) explained:

> A chain of exploitation exists at several levels: metropolitan (developed) countries over peripheral (developing) countries; centers of power in third world countries over peripheral rural areas; and so on down to the village level. In this system, the peripheral areas may gain by getting needed resources, but *the price is domination (by the metropolitan or center areas) over local affairs—curricula, texts, and reforms*, for example. (pp. 308–309, italics added)

Although not writing specifically about Samoa, Allahar's (1989) reference to the Third World in general, aptly applies to the history of Samoa and its current position in the international economy:

> beginning with the voyages of so-called "discovery" and the capture of colonial territory and continuing to the present time, the economic and political structures of the colonial and ex-colonial territories have been *distorted* to meet the needs of metropolitan capitalism. (p. 89)

In addition, Allahar's characterization of Third World economies that have become fully intertwined with the international capitalist economy provides a warning of where Western Samoa may find itself if it continues to entice multinational corporations:

> By tailoring their economies to meeting the needs of advanced ones, peripheral countries become dependent on the advanced for supplies of capital, credit, technology, expertise, and the very market demand that makes possible continued production. Hence local needs and local markets tend to be neglected, for the better part of all economic activity is directed toward external markets and consumers. (p. 90)

But, Western Samoa has the potential to turn the tide away from economic dependency in the world economic system. After all, Samoans thrived for several millennia before contact with the *palagi*. Moreover, their geographical isolation becomes an asset in resisting this dependency because Samoans have the political franchise to pick and choose which influences they accept onto their shores. We can hope that Samoan pride and independence will provide the strength necessary to resist economic domination. Perhaps Samoans can pioneer new models of local autonomy that can be used to help themselves and other countries avoid the cycle of economic dependency and poverty. The future of young Samoans depends on it.

CHAPTER EIGHT

Conclusion: Mead's Samoa in Sociological Perspective

To complete this interdisciplinary examination of the validity of Mead's account, in this concluding chapter I place Mead's Samoa in sociological perspective. With this perspective I round out the assessment as to whether the social structures Mead described in 1920s Ta'u would plausibly have been conducive to a coming of age that was "not a specially difficult" one (Mead, 1928, p. 197). In doing so, I specifically focus on Mead's claim that, with few exceptions, for her informants "adolescence represented no period of crisis or stress, but was instead an orderly developing of a set of slowly maturing interests and activities" (p. 157).

I begin this sociological analysis by backtracking to clarify several concepts bandied about during the Mead–Freeman controversy. Throughout the controversy, references have been made to notions such as "adolescence" as if their definitions and applications were well established. In reality, this notion remains quite controversial in the human development literature. Therefore, to move the Mead–Freeman debate toward a resolution, a thorough analysis of several concepts needs to be undertaken.

ADOLESCENCE AS A STAGE OF LIFE

Any discussion of the concept of adolescence must begin with the recognition that the very notion of "adolescence" is not a culturally universal concept—in some cultures there is no period of "adolescence," as Westerners know it, and there are no words to describe it. Moreover, in cultures where a prolonged period between childhood and adulthood does exist, "adolescence" takes different forms depending on socioeconomic conditions. In short, adolescence is not a universally recognized or institutionalized stage of life; rather, conceptions of it emerge under particular socioeconomic circumstances where a prolonged period between childhood and adulthood can be sustained.

With respect to the idea that a culture may not have a conception of "adolescence" at all, Sprinthall and Collins (1984) stated the following:

> can an adult society ignore something as real as adolescence? That idea may strike you as odd, but such has been the case for adolescence—and for other developmental periods as well. For example, until adults recognized and allowed childhood to emerge, it hardly existed. For long centuries it was thought that as soon as a child reached the age of six or seven, the child was ready to be trained as an adult. Children were considered as little more than midget-sized adults. Except for a tiny proportion of the rich and well-born, they worked alongside adults in the fields, they fought adult wars, they worked in the mines, and with the coming of industrialization they worked from dusk to dawn in the factories. (p. 2)

From this passage, we can see that childhood in preindustrial *Western* culture was closer to the childhood in precontact Samoa as described in chapter 4, than it is to childhood in contemporary Western culture. Thus, childhood has changed in both Western and Samoan culture. Of importance to the Mead–Freeman controversy is the fact that childhood and adolescence can be structured in a variety of ways, and that Western models are neither the only nor the best models to emulate. Moreover, since Mead carried out her study, we have accumulated a considerable amount of knowledge regarding how Western adolescence evolved. For example, as noted by Sprinthall and Collins, the contemporary Western life cycle began to take its present form with the onset of industrialization:

> only in the past 150 years or so have adult Western societies even recognized childhood, the juvenile years from 6 or 7 to 12 or 13, as a special stage of growth. . . . Once childhood was discovered, a whole series of changes followed. . . . What was true for the discovery of childhood in the last century has been the case for adolescence in this century. Only recently have adults in industrialized nations and cultures begun to perceive adolescents' physiological and psychological needs and capabilities as unique, and this perception has created the opportunity to recognize a stage of human growth. . . . In the last half of the twentieth century we are beginning to witness some changes in how adolescents are treated by adult society, similar to the changes experienced by juveniles in the nineteenth century. (p. 3)

Sprinthall and Collins attributed the emergence of adolescence largely to the perceived need to educate the young, as do many other social scientists (see Klein, 1990). Reminiscent of the efforts of missionaries in Samoa described in chapter 4, primary schools were established usually on the initiative of reformers intent on protecting the young from various "evils" associated with urban-industrial influences (cf. Collins, 1979). With primary schools established, the conception of childhood became cemented in peoples' minds. More recently, secondary schools were established, institutionalizing adolescence and cementing peoples' conception of this age period as well.

Although mass education and various laws have had a dramatic impact on the conception of adolescence, there is evidence that the "cement" demarcating adolescence has not yet fully hardened. In fact, it is difficult to firmly define the boundaries of adolescence in contemporary Western societies. For many people, including many social scientists, "adolescence" is taken to correspond to the "teen" years, at least until age 18. According to Montemayor, Adams, and Gullotta (1990), however, in the United States the U.S. legal establishment specifies that adolescence "begins at 16 years of age and ends at 21" (p. 10). In fact, Montemayor et al. devoted their entire volume to trying to understand the transition from childhood to adulthood in contemporary Western societies. Thus, although well institutionalized in Western societies, the demarcation of adolescence is variable, with different observers defining its boundaries in different ways (cf. Gold & Petronio, 1980).

Numerous accounts of the evolution of childhood and the "invention" of adolescence in Western culture are available (e.g., Aries, 1962; Coleman, 1974; Demos & Demos, 1969), so I leave it to the reader to consult these if desired. The point to be taken here, however, is that in light of cultural variation, there are no hard biological "facts" defining adolescence, as there are of puberty.

In spite of these difficulties, the term *adolescence* is used by most social scientists to designate the period between the "biological maturity" produced by puberty and the "social maturity" recognized by an individual's adult culture. Social maturity, however, is not accomplished in some cultures—mainly more affluent cultures—until long after biological maturity has been reached. Under such conditions, the stage of life we call adolescence emerges, and social institutions usually develop to supervise the behavior of those passing through it. Thus, when attempting to discover the primary causes of the behavior of adolescents, we must first gain an understanding of the nature and impact of these social institutions. In addition, what must be emphasized is that biology cannot logically be considered the primary cause of a life stage that is social in origin.

A few words on the expression "coming of age" are also appropriate before proceeding. To come of age in a society is to "qualify" for adulthood because of one's attributes. These attributes may be defined biologically, psychologically, socially, or legally by a society. As was seen earlier, there is no culturally universal standard for the end of adolescence, and hence for coming of age. The term *coming of age* was used by Mead to convey this notion of the transition to adulthood, but she also did not explicitly define the term. Rather, her usage is metaphorical, as is its general usage. The term is primarily heuristic, however, because it draws our attention to the processes of development, rather than to the state of being "adolescent" as opposed to "adult." It also directs us to examine the social institutions that ostensibly *guide* individuals into adulthood.

To conclude this examination of the nature of adolescence, there is no one way for a culture to induct children into adulthood. Equally important is the idea that because adolescence is a cultural phenomenon, we should not expect to discover "instincts" that might guide behavior during this passage. To the contrary, we should expect to find in all well-functioning cultures provisions by which adult generations guide succeeding generations into adulthood. In sociological terms, these provisions are called "institutionalized moratoria" (E. Erikson, 1968). The key feature to be examined is the effectiveness of these institutionalized moratoria in producing social maturity among new members. Accordingly, we now turn to a consideration of adolescence as a social institution. In doing so, we ask questions about the adequacy of that institution in terms of its ability to guide young people through the stage in a relatively stress-free manner. Thus, we have returned to the basic question posed by Mead in *Coming of Age*. Now, however, we have an additional 65 years of theory and research upon which to rely.

INSTITUTIONALIZED MORATORIA AND MEAD'S SAMOA

In reference to the social structures that can emerge to define adolescence, E. Erikson (1980) argued that most cultures provide their new members with some sort of "time out" between childhood and adulthood. This usually involves a moratorium from adult responsibilities, during which they can take time to develop their adult identity. In his words: "Societies offer, as individuals require, a more or less sanctioned intermediary period between childhood and adulthood, institutionalized psychosocial moratoria, during which a lasting pattern of 'inner identity' is scheduled for relative completion" (p. 119).

There is also usually some form of structure or guidance given the young person (hence the term *institutionalized*). Together with a freedom to experiment with various roles (hence the term *moratorium*), young people can explore themselves and their world without being expected to carry permanent responsibilities and commitments. In E. Erikson's (1968) words, "these moratoria coincide with apprenticeships and adventures that are in line with society's values" (p. 157). Apprenticeships include schooling and forms of "service" (military or voluntary, like the Peace Corps), whereas adventures can take various forms, including travel (the *Wanderschaft*), or just "dropping out." In societies that institutionalize them, all of these forms of activity are socially recognized as legitimate for young people, as opposed to, say, children or the elderly.

Mead's description of the period granted to both males and females on Ta'u corresponds in several ways to this description. For males, as we saw earlier, "service" (*tautua*) in the *'aumaga* was well institutionalized. This experience provided some opportunity for the type of role experimentation that would contribute to the "lasting pattern of 'inner identity' " about which Erikson writes. Moreover, the *'aumaga* constituted a form of apprenticeship (as did the *aualuma* in precontact culture) that would have provided "a moratorium characterized by defined duties and sanctioned competitions as well as special license" (E. Erikson, 1968, p. 185). In precontact Samoa, these apprenticeships would have been complemented with elaborate initiations and welcoming ceremonies (see chapter 4).

Evidence of the "special license" of the moratorium (temporary freedoms not granted adults or children) can be found throughout Mead's account. For example, as was seen earlier, she described how many of her informants were permitted to forestall full adult responsibilities with the defensive statement, "I am but young." She also described the "selective playfulness" among the young that was socially recognized (implicitly endorsed) by many adults with respect to sexual experimentation. In addition, both males and females participated in the *malaga* (visiting parties to other islands and villages), as well as in expeditions to gather food and craft supplies, so something equivalent to a *Wanderschaft* was available.

Characteristics of Effective Institutionalized Moratoria

With the concepts of adolescence and institutionalized moratoria clarified, I now specify four social conditions that should be present to minimize the possibility that young people will experience difficulties as they pass from childhood to adulthood in a given society. For each of these four conditions, I then examine Mead's Samoa to determine whether these would have been met.

Social Organization

For adolescence to be relatively stress free, a society should be adequately organized so that this transitional stage is marked by a *continuity* between childhood and adulthood (cf. Benedict, 1938). Thus, a learning of the basis of adult roles should be accomplished during this stage, and this learning should be a logical extension of what was learned as a child. Similarly, the acquisition of adult responsibilities should take place in a fashion that does not tax the individual's capacity to exercise those responsibilities. In cultures where these conditions are not met, we should expect to find adolescents encountering difficulties both in terms of their day-to-day behavior as "adolescents" and in terms of their ability to project themselves into the future. Cultures that are *socially disorganized* lack integration among their institutions. For example, there may be a lack of coordination between the family and the educational system, or between the educational system and the adult world of work. Under these circumstances, adolescents are left to deal individually with conflicting expectations when moving from one institutional setting to another. The experience of this type of conflict can create psychological confusion and interfere with the development of attributes defined as appropriate to adulthood in a given culture. Ideally, institutions work in unison so that as individuals move from one setting to another, they do not encounter serious disjunctures of experience: Thought and behavior patterns that prevail in one setting should be more or less appropriate in another setting, especially if one setting is supposed to constitute a preparation for another (as adolescence ostensibly is to adulthood).

Social Organization in Mead's Samoa. From Mead's descriptions of the structural arrangements governing adolescence and the transition to adulthood, there is little doubt that they provided adequate guidance for most young Samoans. Several factors should have ensured this.

First, for both males and females, learning took place gradually, and young people moved on to more complex tasks as they were able. There was apparently little pressure to take on tasks for which the person was not prepared.

Second, there was evidently no shame in being "slow" at something because of the cultural norm that set group activities at the pace of the slowest person. Moreover, undue "precocity" was negatively sanctioned, especially among the young, so there was little point in someone trying to outdo someone else. Mead (1928) noted, however, that when she conducted her study the norms regarding precocity were diminishing in importance, primarily because of the influence of the educational system. In her words: "Even the stern attitude formerly taken by the adults towards precocity has now been subdued, for what is a sin at home becomes a virtue at school" (p. 276).

Third, individuals of all ages would have felt few ambiguities regarding what was expected of them because of the stable and explicit definition of roles.

These factors would have combined to not only reduce competition, but also to minimize the extent to which individuals compared themselves to some ideal standard of accomplishment (the exception to this was the individual aspiring to a *matai* title). Thus, each individual would have known where they stood in relation to community roles, and membership in village groups was well defined.

It is worth arguing that these cultural arrangements make great sense in an island-bound culture where there are limited resources and quite literally no where else to go. In other words, there is little point in encouraging innovation when everyone has enough to eat and where there are few material resources to be accumulated. Moreover, the discouragement of innovation and precocity would ensure cultural stability—a "resource" apparently much more important to an island-bound culture than would be any material comforts accruing from innovation. The Western obsession with "achievement" and "striving" has led to world domination, but Westerners are not necessarily any more fulfilled than were Samoans before Western contact. Indeed, it appears that the Western middle-class "future orientation" has now been internalized by many Samoans, convincing them that they must be dissatisfied with their present circumstances in order to be happy in the future.

These features of Samoan social organization would account in part for the "casual attitude" reported by Mead. Apparently there was simply little pressure on most young people to hurry, to be extremely productive, or to "prove themselves" through innovation.

Role Induction

Effective institutionalized moratoria should provide some sort of role induction. By this I mean that *role learning* in adolescence should constitute a preparation for adult roles and a basis on which the young recruit is *welcomed* into an adult community. To the extent that roles played in adolescence are irrelevant to, or in conflict with, adult roles, the transition to adulthood can become problematic. This may be manifested either as part of a lack of welcoming from the adult community or a "refusal" of the recruit.

In Western culture it can be argued that much of the role learning and behavior of adolescence is irrelevant to adulthood. Many adolescents have few work or family responsibilities, they must function with the low status associated with the student role, and they have few practicable choices about their present lifestyle. Accordingly, the youth cultures and subcultures now common in many Western societies can be seen to have emerged to fill the "voids" of adolescence—particularly as they represent resentment about these voids and a diffuse rebellion against "adult authority" (cf. Brake, 1985). Adults, when reacting to these subcultures (particularly gangs), often exhibit fear or contempt, driving a further wedge between the adolescent and the adult community. This cycle of resentment appears to be a product of the force prolonging the period of youth and increasing the disenfranchisement of young people. In more regulated societies, the differences between the adolescent world and the adult world are minimized.

Role Induction in Mead's Samoa. By the time Mead did her study in the 1920s, many structures associated with the induction into adult roles had been altered or eliminated by the missionaries. For example, the requirement of tattooing for membership in the *'aumaga* had apparently been taboo for two generations (Mead, 1928, p. 267) and ceremonies seldom marked induction into the *aualuma* (p. 275). Nevertheless, it does not appear that the general "role induction ethos" of Samoan culture had yet been fully compromised by missionary influences.

For young males, preparation for adult roles was still evident. By their early teens they were skilled in the basics of adult work activities and during their "youth" they worked alongside adults in the *'aumaga*, which would eventually introduce them to full adult status. In addition, the license to remain single enabled those who did so to take life more casually than did those who married early, especially in comparison to those married men aspiring to a *matai* title.

Young females encountered less structure, but they too learned most of the skills of adulthood during their childhood. The period between puberty and their early 20s permitted them the time to hone these skills and to prepare psychologically for their adult roles, apparently at a relatively leisurely pace. In particular, fine mat weaving and other craft skills were developed during this period. Upon marriage, females were considered "adults" and were apparently well prepared for this status. Thus, because of their skills preparation, and the recognition of this by their community, the moratorium they experienced before marriage was more "socioemotional" than "instrumental" in terms of preparation for adulthood.

Social Control Mechanisms

The logical extension of the first condition and the requisite of the second condition, is that social control mechanisms must be in place to ensure that behavior is regulated. These mechanisms should be structured to ensure that roles are meaningfully integrated into a social order and division of labor. Social control mechanisms can be defined as patterns of reward and punishment that direct behavior in certain ways. These contingency patterns make clear what behavior is appropriate or desirable, as defined by cultural norms and values.

When these mechanisms do not exist, it is unclear to individuals, particularly new individuals attempting to learn the requirements of the culture, what they should do to act appropriately. In other words, individuals can become confused without a well-regulated social environment. Moreover, it is not enough that norms simply *exist* regarding what is right and wrong; these norms must also be *enforced* by human agents. If they are not enforced, the norms become meaningless over time as fewer individuals internalize or respect them. When adolescents are confronted with poorly constructed and ineffectively executed social control mechanisms, they can experience, among other things: (a) confusion regarding how to behave; (b) a lack of attachment to, and resentment toward, individuals subjecting them to these mechanisms; (c) an absence of moral conviction; (d) a lack of meaning in their lives; and (e) a sense of confusion about whom they are and where they are headed as individuals. All these have become central "symptoms" of the turbulent and stressful forms of adolescence found in Western societies. Indeed, these symptoms are so commonplace that they are now considered "normal" characteristics of adolescence in many Western societies. Yet, for the young, they can be painful and can stimulate adolescents to organize themselves into subcultures (especially gangs) in an attempt to give their lives meaning and structure.

Social Control Mechanisms in Mead's Samoa.

The institutions confronting Samoan youth were well-structured and well-maintained, partly because of the effectiveness of social control mechanisms. As discussed earlier, these social control mechanisms included the day-to-day interpersonal normative pressures associated with living in a small community where very little takes place in private; the nature of the Samoan *aiga* and the requirement of loyalty and duty to it from its members; the *ifoga* (forgiveness ceremony); the *tautua* (service to the *matai*); the allotment of fines and punishments by the *fono* to the offender *as well as* his or her *aiga* and *matai*; and banishment or execution if the offense were severe enough. Apparently, punishments were immediate (cf. Gerber, 1975), and were perceived by most violators to be justified (cf. Holmes, 1987, p. 106). Moreover, the apprehension regarding shaming oneself—and by implication one's *aiga*—would have been an effective approach/avoidance motivator, particularly because of the lack of privacy. A "shame-oriented" culture can be much more orderly than a "guilt-oriented" one, because shame is an overt, collective phenomenon, whereas guilt is a covert, personal one (cf. Gerber, 1975).

Deviance

Finally, stemming from Condition 3, social control mechanisms must be effective in dealing with various types of deviant behavior. One type of deviant behavior can be seen as a "normal" by-product of social organization. This type of behavior is normal because it stems from the fact that when rules (norms) exist, violations will eventually ensue. In other words, if there is no rule with respect to a behavior, there can be no violation. For example, if there is no rule prohibiting premarital sexual behavior, there will be no deviance in this regard, and no one will be labelled as this type of "deviant" (cf. K. Erikson, 1966). A second type of deviance constitutes a by-product of particular types of culture. In this view, "culturally endemic" deviance results from poorly integrated reinforcement patterns that cater to our base capacities as "human animals," such as selfishness and egocentrism. When a culture rewards these animal capacities, they may take prominence over other human capacities such as loyalty, sharing, and duty. Yet another type of deviance is the result of a cultural incapacity to impose its normative pattern on members with idiosyncratic "temperaments" (e.g., as described by Mead concerning some of her "deviants"). Each of these sources of deviance constitute challenges for all cultures to establish mechanisms that ensure relatively harmonious day-to-day community functioning.

Deviance in Mead's Samoa.

The previously mentioned social control mechanisms should have discouraged deviant behavior in the first instance, and remedied any deviance-related damage to community relations in the second instance. With respect to the incidence of deviance in precontact culture, it was noted earlier that the missionaries were very thorough in documenting the "evils" they perceived. Accordingly, had serious problems existed among the young, such as suicide and delinquency, the missionaries most certainly would have mentioned them and set out a strategy to deal with them. No mention was made of a serious problem of deviance among the young, however.

As a thorough researcher, Mead extensively documented the deviance in the three villages she studied. In fact, Mead's Appendix IV deals specifically with the topic of "The Mentally Defective and the Mentally Diseased," and her chapter XI deals with the question of "The Girl in Conflict." Moreover, references are made throughout the book to the concerns associated with "deviance and social control" issues. For example, in reference to the issue of "crime" and the impact of Western justice on precontact Samoan justice, Mead wrote that

> the Samoan is not in the habit of committing many of the crimes listed in our legal code. He steals and is fined by the government as he was formerly fined by the village. But he comes into very slight conflict with the central authorities. He is too accustomed to taboos to mind a quarantine prohibition which parades under the same guise; too accustomed to the exactions of his relations to fret under the small taxation demands of the government. (p. 276)

In other words, most Samoans were still influenced by their socialization into precontact culture, and were motivated by conditioning from that culture rather than from a fear of legal action from government officials.

The deviance that Mead detailed in her chapter entitled "The Girl in Conflict," has been the object of criticism by Freeman. As noted in chapter 2, most of the deviance discussed by Mead was minor, and would not likely have resulted in criminal indictment in Western societies. Rather, this deviance constituted either "serious conflict" or "serious deviation from group standards." Readers will recall that two cases of serious conflict were recorded, both of which had begun before puberty. In addition, three cases of "serious deviation from group standards" were noted, having been precipitated by an awareness of alternative adult roles stemming from missionary/Western influence; and two additional cases of minor or potential deviance were recorded: one potential delinquent, and one "delinquent in the making" (p.182). In Mead's estimation, however, this deviance "in temperament or in conduct . . . [was] . . . in many cases . . . only charged with possibilities of conflict and actually had no painful results" (p. 158). The remainder of her informants, "showed a surprising uniformity of knowledge, skill and attitude, and presented a picture of orderly, regular development in a flexible, but strictly delimited, environment" (p. 184).

With respect to the storm and stress issue, if we interpret the aforementioned conflicts to represent symptoms of storm and stress, then the following comparisons can be made with recent North American research. Of the 25 postpuberty informants, three could be said to have experienced some minor difficulties in what is now called "identity formation" by human development researchers (establishing "a lasting pattern of 'inner identity'," as quoted from Erikson), although two experienced more serious and chronic "delinquency." Thus, 12% (3 out of 25) of Mead's informants might have experienced difficulties in identity formation, compared with 50% to 70 % of contemporary Canadian and U.S. young people (e.g., Côté, 1986; Marcia, 1980). An additional 8% (2 out of 25) engaged in some form of "delinquent behavior," compared with up to 80% for U.S. young people (Gold & Petronia, 1980). Thus, even with this generous stretch of the notion of storm and stress, it appears that Mead was reasonable in her conclusion that most of her informants experienced a coming of age that was "not a specially difficult" period and that adolescence itself did not represent an inevitable "period of crisis or stress."

If the four sociological conditions just listed are met, the young person should find some sort of order in roles, a predictability in outcome, and a sufficient reward for adhering to social conventions. From this account, there are ample grounds to conclude sociologically that coming of age in "Mead's Samoa" occurred as part of a relatively smooth passage from childhood to adulthood for the majority of her informants, and that any symptoms of adolescent storm and stress would have been less common than is now the case in Western culture. The charge has been made, however, that coming of age was *experienced* as stressful by many young Samoans (see Freeman, 1983, chapter 17). One way to assess the experiential aspect of this issue is to focus on one of the primary differences between adolescence in Mead's Samoa and adolescence in Western cultures, namely, the amount of choice available to the young person.

THE ISSUE OF CHOICE

One final issue involves something that might have troubled those readers who began this book skeptical that social conditions in Mead's Samoa would have been experienced by most young people as benign. The issue in question is having the "right" to choose one's destiny.

In comparison to Western societies, in Mead's Samoa young people had far less choice regarding the specific content of their future adult identities. Mead noted that this had the effect of eliminating much conflict from their lives: They did not have to choose among competing religions; they did not have to choose among competing political philosophies; and they did not have to choose from among a bewildering array of adult occupations. In contrast, Mead noted there was a virtual requirement in the United States that young people choose for themselves among the myriad of religious, political, and occupational

THE ISSUE OF CHOICE 161

options. In Mead's (1928) words: "The principal causes of our adolescents' difficulty are the presence of conflicting standards and the belief that every individual should make his or her own choices, coupled with a feeling that choice is an important matter" (pp. 234-235).

In contrast, in Mead's Samoa:

> The growing child is faced by a smaller dilemma than that which confronts the American-born child of European parentage. The gap between parents and children is narrow and painless, showing few of the unfortunate aspects usually present in a period of transition. . . . essentially the children are still growing up in a homogeneous community with a uniform set of ideals and aspirations. (p. 273)

Aptly noting that "the need for choice [is] the forerunner of conflict" (p. 202), Mead predicted that the introduction of choice into the lives of Samoans coming of age was going to have dramatic implications. As we have seen, Mead appears to have been right. One reason why the issue of choice had not yet affected most of her informants was the lack of exposure to many Western institutions. Even the pastors' village boarding schools were not yet affecting the world view of most of Mead's young informants, and they were "likely to pass through the school essentially unchanged [in their] fundamental view of life" (p. 170). In contrast, those "who left their village and spent several years in the boarding school under the tutelage of white teachers were enormously influenced" (p. 170), largely in terms of adopting nontraditional adult identities based on a career as a nurse or a pastor's wife. Mead saw an end to this conflict-free situation, however, as is evident in the following passage:

> while religion itself offered little field for conflict, the institutions promoted by religion might act as stimuli to new choices and when sufficiently reinforced by other conditions might produce a type of girl who deviated markedly from her companions. That the majority of Samoan girls are still unaffected by these influences and pursue uncritically the traditional mode of life is simply a testimony to the resistance of the native culture, which in its present slightly Europeanized state, is replete with easy solutions for all conflicts; and to the apparent fact that adolescent girls in Samoa do not generate their own conflicts, but require a vigorous stimulus to produce them. (pp. 170-171)

The exceptions were the "upward deviants," who had been stimulated to "deviate" by their exposure to Western education. They wished

> to exercise more choice than is traditionally permissible, and . . . in making their choices, come to unconventional and bizarre solutions. The un-

traditional choices which are encouraged by the educational system inaugurated by the missionaries are education and the pursuit of a career and marriage outside of the local group . . . a self-conscious evaluation of existence, and the consequent making of self-conscious choices. All of these make for increased specialization, increased sophistication, greater emphasis upon individuality, where an individual makes a conscious choice between alternate or opposing lines of conduct. (p. 171)

Mead argued that the lack of choice was not experienced as oppressive or as a misfortune by most of her informants, except for the few who were "temperamentally" unsuited for such circumstances. Rather, she saw several mechanisms in Samoan culture that socialized people to accept this situation and to not view it as negative. For example, *children* were "urged to learn, urged to behave, urged to work, but they [were] not urged to hasten in choices which they make themselves" (p. 231), and *adolescents* were not "pressed to make momentous decisions which would spoil part of their fun in life" (p. 232). In addition, by "the time she reaches puberty the Samoan girl has *learned to subordinate choice* in the selection of friends or lovers to an observance of certain categories" (p. 210, italics added). Many Westerners balk at the idea of restrictions of choice, reflexively viewing such circumstances with disdain. It is worth examining this issue further, however, because of the implications of the insights yielded from such an examination. For example, even at the level of common sense, it is easy to realize that an unbridled freedom of choice does not necessarily lead to happiness and fulfillment. As Mead pointed out, too much choice can be a source of unhappiness and confusion, especially when there is little guidance in choice making or when the choices are illusory.

I believe that when we are sensitive to the cultural requirements of Mead's Samoa, it would be the presence of choice, not the absence of choice, that would be incongruent with cultural identities. Further, the presence of choice is likely responsible for much of the identity confusion and stress and storm that can now be witnessed in contemporary Samoa. Indeed, this preoccupation with choice seems to be very much a Western, Protestant, and 20th-century phenomenon (cf. Baumeister, 1986; Mead, 1928, pp. 232–233, located the Western concern over individual choice with the Reformation and the rise to prominence of Protestantism). There is ample evidence from North American studies that excessive occupational and ideological choice among those moving from childhood to adulthood can itself be stressful. This is partially because choice is often an illusion for many people—they simply cannot "be" or do many of the things to which they have aspired because of a lack of opportunity.

In other words, identity confusion often stems from "over-choice" in a situation where the actual range of choice and control over choices is quite limited, especially for those from the less privileged groups of a society (Côté & Levine, 1987). Indeed, the introduction of choice and the competition that goes along with it has been found to be very stressful for many contemporary young Samoans, especially those raised under more traditional circumstances (cf. Baldauf & Ayabe, 1977).

Obviously, this book is not the place to exhaust the monumental issue of "freedom of choice." It is an issue over which revolutions erupt and civil wars are waged. What can be said in the context of the coming-of-age issue, however, is that in our desire for absolute freedom of choice in the West, we have forgotten that humans are not genetically programmed for making specific choices during their lives. Rather, exercising choice is a capacity to be acquired on the basis of experiences within social institutions. What we have forgotten in Western culture is that young people need guidance in developing the capacity to make choices that are personally and socially responsible. This guidance should be present in the institutions that ostensibly prepare the young for adulthood. In the West, we have failed many young people in this regard; in imitating Western culture, contemporary Samoan culture is making the same mistake.

The significance of this problem can be put into context when we examine the relationship between mental health and social organization. Seligman (1988), for example, argued that the "rampant individualism" of U.S. society, along with a decline in commitment to common social institutions, has led to high rates of emotional depression in the span of the last two generations. In fact, it is estimated that "today's young [American] people [are] about 10 times as likely to be depressed as were their parents and grandparents" (pp. 50–52). Seligman attributed this "epidemic of depression" to the tremendous expectations placed on the "self" to be both the source and the "architect" of most forms of gratification— to "decide, prefer, plan, and choose [its] own course of action" (p. 52). However, this selfishness "need not lead to depression as long as we can fall back on large institutions" (p. 55). I would add, following Mead's conclusion, that mental health problems can be averted if "coming-of-age institutions" are properly and benignly structured in the first place to give the individual the capacity to function in an autonomous, yet "socially responsible," manner (cf. Côté & Levine, 1987). With "democracy" breaking out globally during the 1990s, we must be careful to balance the benefits associated with the right to choose, against possible liabilities. These liabilities are created when people do not have the psychological capacities to make complex choices, or when the opportunities are actually illusory, and no amount of effort or capability will produce the desired goal.

LIMITATIONS OF THE PRESENT STUDY

Before concluding, it is appropriate for me to reflect upon the limitations of the study described in this book. The primary limitation of this study of Mead's coming-of-age thesis and the Mead–Freeman controversy is that direct data pertaining to Samoan youth are sparse, particularly information pertaining to precontact Samoa. The social history constructed in chapters 4 and 5, therefore, was reliant on secondary sources and much of the information was descriptive, so it was often left to the investigator and reader to draw their own conclusions regarding coming-of-age issues. These secondary sources were often European observers who may not have appreciated the fullness and richness of Samoan culture. Moreover, as male observers they may not have appreciated the role of females in Samoan culture, so a number of important details may have been lost. With these problems in mind, as much as possible I used information that was verified by other sources, especially scholars from Samoa (e.g., Meleisea) and women scholars (e.g., Schoeffel). Fortunately, some records were kept and many of the practices survived well into this century.

The same limitation of sparse data applies to information examined in chapters 2 and 3. Much of the information examined was not collected with the primary purpose of studying coming-of-age processes, and it is not possible to go back and repeat Mead's study of the 1920s (although Holmes' restudy during the 1950s supported Mead's thesis). Consequently, it was necessary to rely on inference, and to piece the information together much like is done in a legal investigation. The comparisons made of Freeman's evidence with evidence from other parts of Samoa and from other countries were enlightening, but one is still dealing with inferences and issues of plausibility rather than definitive conclusions that might be drawn from controlled experiments. However, to compensate for these limitations, evidence was examined in light of state-of-the-art theory and research that was not available to Mead, and that Freeman has not used.

These are the approaches taken when dealing with legal issues, so early in this book I asked readers to take the role of "jurists" in evaluating the evidence and arguments. Only when enough evidence had been compiled on which to base a sound judgment, could a verdict on the matter be made. Of course, as with legal decisions, the verdict passed in this case only stands as valid in terms of "reasonable doubt." But, as I argued, there is reasonable doubt with all of Freeman's evidence and arguments with respect to Mead's coming-of-age thesis, regardless of which disciplinary perspective is taken. As demonstrated, each piece of evidence examined was either discounted as misleading or interpreted in ways that do not refute Mead's thesis.

FUTURE RESEARCH

It was just noted that a limitation of this study is that it had to deal with a paucity of data about Samoan youth. Although more information is now available about youth in contemporary Samoa, researchers should take up the task of providing effective documentation of current circumstances so that another controversy does not arise that presents as many ambiguities as does the Mead–Freeman controversy (see Baker, Hanna, & Baker, 1986, for such efforts). These research efforts should also be directed at cataloguing youth problems and making recommendations to help rectify them. Thus, aside from descriptive documentation, there is some urgency that research efforts be directed at identifying what can be done to buffer Samoan youth from the impact of Western culture and from international capitalism. In doing so, this research should look to the strengths of Samoan culture and character, as well as to weaknesses created by social change that leaves young Samoans vulnerable as they struggle to come of age. For a model, see Reid's (1990) study of moral reasoning. Note her identification of the sociocentrism institutionalized in the Samoan *fono* and the postconventional reasoning associated with methods of dispute resolution. Researchers should also consult Flay, Bull, and Tamahori (1976) for a treatment of how to develop questionnaires appropriate for Polynesian respondents.

In terms of a major research thrust, one particularly promising line of study has been developed by North American researchers, but before that line of study can realize its potential, the issue of cross-cultural validity must be confronted, as I explain later. This line of research focuses on "identity formation" as a central issue in adolescent development. Its promise lies with its "psychosocial" perspective, namely, the potential to incorporate psychological concepts with sociocultural ones (cf. Côté, 1993).

The line of research in question is referred to as the "identity status paradigm" (e.g., Marcia, 1964, 1976, 1980) and is based on Erik Erikson's (e.g., 1956, 1958, 1963) theory of ego development. Central to this paradigm is the issue of conscious choice made by individuals regarding life plans and belief systems. The period during which individuals choose among career and belief options is postulated by identity status researchers to constitute a form of identity crisis. This identity crisis, or period of conscious decision making, can last for years. Even so, it is postulated that experiencing a period of conscious decision making is in general psychologically beneficial for the individual. Evidence to support this claim links the crisis with certain personality characteristics like autonomy, internal control, cultural sophistication, and the need for achievement (cf. Côté & Levine, 1988a). Moreover, on the basis of the belief that experiencing this identity crisis is good for the individual, these researchers also recommend that everyone be encouraged to experience it, and therefore "achieve" an identity.

Those who do not go through this period of conscious decision making are said by identity status researchers to have avoided the identity crisis and "foreclosed" on their adult identity formation. That is, they do not experience this identity crisis and they uncritically accept family- or community-conferred roles. It is further assumed that people who foreclose on their identity formation suffer certain deficits, including a more limited sense of "ego identity" (a sense of temporal-spatial continuity).

Applying this paradigm to Mead's informants, it follows that nearly all would be classified as "identity foreclosed." Accordingly, they would be viewed by identity status researchers as less "advanced" in terms of ego development than American youths. A problem with this form of analysis, then, is that it is based on the assumption that ego development is greater if an individual experiences a Western-style "identity crisis," defined in terms of the activity of consciously exploring options regarding work roles and beliefs. But, the ability to engage in choice behavior was limited by Samoan culture, as it is by most non-Western cultures. Moreover, Mead's "deviants" would likely be classified as "identity moratorium" in the case of the upward delinquents, and as "identity diffusions" in the case of the downward delinquents. ("Identity moratorium" describes the process of deciding about which adult commitments to make, while "identity diffusion" involves avoiding both the decision making and commitments.)

With respect to the cross-cultural relevance of the identity status paradigm, Marcia (1976) argued that "when the statuses are applied in other cultures they will take on different significance, according to a particular culture. For example, Foreclosure will have a different meaning . . . in a society where roles are prescribed and an 'identity crisis' is not the norm" (p. 117). Marcia referred to this type of identity formation as "developmental Foreclosure" because such individuals have not "been exposed to an identity-challenging environment" (p. 118). In spite of this declaration, very little research or theory has dealt with the ego development that takes place for "developmental foreclosures."

Thus, it can be argued that the issue of the cross-cultural implications of the identity statuses is undertheorized and underresearched with this paradigm. Although it highlights the cultural relativity of the issue of choice, the identity status paradigm does not provide the theoretical concepts with which to understand the types of ego development that might be accomplished by people in cultures that are not choice- and self-oriented. In contrast, in Erikson's general theory, people of all cultures can develop a strong sense of ego identity based on role validation and community integration, especially when there is a lack of ambiguity regarding beliefs (cf. Côté & Levine, 1988b). In speaking to this general issue, E. Erikson (1963) wrote that

> even the most 'savage' culture must strive for what we . . . call a 'strong ego' in its majority or at least in its dominant minority—i.e., an individual core firm and flexible enough to reconcile the necessary contradictions in any human organization, . . . and above all to emerge from . . . infancy with a sense of identity and an idea of integrity. (pp. 185–186)

On the question of cross-cultural applications of the identity status paradigm, precontact Samoan culture would be called a "foreclosed society" (Marcia, 1976). As just argued, however, individuals in all societies should have a generally good sense of ego identity if they are effectively supported by their community. In foreclosed societies they may not have the autonomous form of identity referred to by identity status researchers, but it would be culturally inappropriate for them to have such an autonomous identity. So, herein lies the challenge to identity status researchers—to adapt the paradigm to be sensitive to forms of psychosocial development that occur in other cultures.

Instead of a form of ego development that helps people function in middle-class urban environments, individuals in foreclosed societies likely have other strengths that help them function and continue to mature in those environments (see Mageo, 1989, for a linguistic-anthropological study of the structure self and identity in Samoa). For example, one would expect advanced forms of ego development to be stimulated as Samoans passed through Erikson's last three psychosocial stages of intimacy, generativity, and integrity. All three of these stages appear to have been well-institutionalized in precontact culture, and in Erikson's view, the better institutionalized a stage, the more it nurtures ego development.

In fact, an argument can be made that these stages—particularly generativity and integrity—were more thoroughly institutionalized in precontact Samoan culture than in contemporary Western culture, where the glorification of choice has led to a "selfishness," as opposed to Samoan "otherness." In Erikson's stage hierarchy, fulfilment is derived by becoming less selfish, rather than more selfish, as one encounters the stages of intimacy, generativity, and integrity. Because they were well-institutionalized, one would expect people to enter these stages earlier and to move through them more quickly, thereby accruing and using their associated strengths for a greater portion of their lives. In terms of specific institutions, the *'aumaga* and the *fono*, for example, unquestionably stimulated various skills associated with ego mastery (cf. Reid, 1990).

Nevertheless, some Western readers may still feel that the relative lack of choice may be stress-producing and an impediment to development for many adolescents in any culture. This may be so for some individuals, but if we use Mead's study as a reference point it is unlikely that more than a handful of her informants were bothered by it (i.e., the three "upward deviants"). Certainly, there is ample evidence that Samoan childrearing techniques and cultural conditioning prepared those coming of age for this, by effecting a high degree of conformity-oriented behavior and a low degree of individuated or "creative" behavior (e.g., Holmes, 1987). Indeed, similar childrearing techniques prevail in other South Pacific cultures. As Crocombe (1989) argued, in many South Pacific cultures, typical socialization is such that individuals "have been conditioned by their cultures not to 'push' themselves . . . much of their cultural conditioning has taught them not to innovate, not to work out their own destiny, not to strive to improve their position" (p. 40).

The implications of this cultural conditioning for the identity status paradigm emerge from one of the few cross-cultural studies conducted using that paradigm. In New Zealand, Chapman and Nichols (1976) assessed *occupational* identity status among Maori (Polynesian) and Pakeha (European descent) high school students. They found that more Maori youth were "identity diffused" ("not committed to an occupation and had not seriously contemplated possible occupations"), although more Pakeha youth were "identity achieved" (p. 61). These findings were interpreted with the assertion that "Maoris typically see occupation as a means of providing the necessities of life and of making it possible to extend hospitality while Pakeha more commonly see occupations as a vehicle to and index of success" (p. 65). Noting differing cultural meanings regarding choice and commitment around occupation, Chapman and Nichols argued that the "greater frequency of identity diffusion among Maoris may indicate that this is an adaptive, role-appropriate status for Maoris" (p. 69).

The Chapman and Nichols study suggests that several questions need to be addressed by future research. For example, just why might something like "identity diffusion" be adaptive for one group and not another? And, how might individuals from such groups develop strengths independent of, or in spite of, such conditions. In addition, a specific research question that can be pursued in New Zealand (and other Western countries with indigenous populations) is to identify what replaces *conscious decision making* as a source of ego development in Maori culture. One avenue to explore in this regard is the *guidance* the culture may provide the young that makes conscious decision making unnecessary as a stimulus to development. Interestingly, this brings us back to Mead's concern over conflict and choice, discussed earlier.

A challenge for future research is to adapt theories developed by

Western researchers to other cultures in the sense of determining the psychosocial strengths associated with various forms of cultural conditioning. The past undertheorizing about cultural differences in socialization and human development is an impediment to this task. One starting point is to recognize macro, political-economy perspectives, such as those discussed in the last chapter, and to link them with micro, sociopsychological, and developmental concerns discussed in this chapter.

An issue for this research to deal with pertains to the problem of identity in a world in which there is a declining respect for authority and tradition. This decline of authority has been ongoing in Western societies for some time, but as it has taken root in places like Samoa, its repercussions have been sudden and dramatic. Accordingly, societies like Western Samoa provide a site in which to study change that has accelerated in a very short time, in relation to the less rapid change in Western societies. Comparisons of these changes should shed light on problems faced in both types of societies. For example, it is clear from the study just presented that as a culture grapples with its identity, those attempting to come of age in that culture also grapple with their identity. Thus, change that is too rapid and unregulated is detrimental to identity formation processes. Without a clear sense of identity, humans have a very difficult time functioning as social beings. The dawning global civilization engendered by international capitalism must deal with this, but to do so a sound understanding is needed of the relationship between culture and identity.

These are the types of challenges and issues that inspired Margaret Mead.

References

Adams, G. R., Montemayor, R., & Gullotta, T. P. (1990). *Biology of adolescent behavior and development*. Newbury Park, CA: Sage.
Aiavao, U. (1991, July). Slow suicides in Samoa. *Pacific Islands Monthly*, p. 37.
Ala'ailima, F. (1984). Review of Margaret Mead and Samoa. *Pacific Studies*, 7, 91-92.
Allahar, A. L. (1989). *Sociology and the periphery: Theories & issues*. Toronto: Garamond Press.
Appell, G. N. (1984). Freeman's refutation of Mead's *Coming of Age in Samoa*: The implications for anthropological inquiry. *The Eastern Anthropologist*, 37, 183-214.
Appell, G. N., & Madan, T. N. (1988). *Choice and morality in anthropological perspective: Essays in honor of Derek Freeman*. Albany, NY: State University of New York Press.
Aries, P. (1962). *Centuries of childhood: A social history of the family*. New York: Random House.
Badock, C. R. (1983). *Margaret Mead and Samoa: The making and unmaking of an anthropological myth* [review]. *British Journal of Sociology*, 34, 606-607.
Baker, P. T, Hanna, J. M., & Baker, T. S. (1986). *The changing Samoans: Behavior and health in transition*. New York: Oxford University Press.
Baker, T. S. (1984). *Margaret Mead and Samoa: The making and unmaking of an anthropological myth* [review]. *Human Biology*, 56(2), 402-404.
Baldauf, R. B., Jr., & Ayabe, H. I. (1977). Acculturation and educational achievement in American Samoan adolescence. *Journal of Cross-Cultural Psychology*, 8, 241-255.
Ballantine, J. H. (1989). *The sociology of education: A systematic analysis* (2nd ed.). Englewood Cliffs, NJ: Prentice-Hall.
Barnouw, V. (1983). Coming to print in Samoa: Mead and Freeman. *The Journal of Psychoanalytic Anthropology*, 6, 425-433.
Baumeister, R. F. (1986). *Identity: Cultural change and the struggle for self*. New York: Oxford University Press.
Benedict, R. (1938). Continuities and discontinuities in cultural conditioning. *Psychiatry*, 1, 161-167.
Beneteau, R. (1988, Winter). Trends in suicide. *Canadian Social Trends*, 22-24.

Berg, I. (1970). *Education and jobs: The great training robbery.* New York: Praeger.
Bernard, R. H. (1988). *Research methods in cultural anthropology.* Newbury Park, CA: Sage.
Bock, P. K. (1983). The Samoan puberty blues. *Journal of Anthropological Research, 39,* 336–340.
Bowles, J. R. (1985). Suicide and attempted suicide in contemporary Western Samoa. In F. X. Hezel, D. H. Rubenstein, & G. H. White (Eds.), *Culture, youth and suicide in the Pacific: Papers from an East-West Center Conference* (pp. 15–35). Honolulu: Center for Asian and Pacific Studies, University of Hawaii at Manoa.
Brady, I. (Ed.). (1983). Special section: Speaking in the name of the real: Freeman and Mead on Samoa. *American Anthropologist, 85,* 908–947.
Brady, I. (1991). The Samoan reader: Last word or lost horizon? *Current Anthropology, 32,* 497–500.
Brake, M. (1985). *Comparative youth culture: The sociology of youth cultures and youth subcultures in America, Britain and Canada.* London: Routledge & Kegan Paul.
Brewer, W. K. (1975). *Armed with the spirit: Missionary experiences in Samoa.* Provo, UT: Young House. (Original work published 1930)
Broad, W., & Wade, N. (1982). *Betrayers of the truth: Fraud and deceit in the halls of science.* New York: Simon & Schuster.
Brooks-Gunn, J., & Reiter, E. O. (1990). The role of pubertal processes. In S. S. Feldman & G. R. Elliott (Eds.), *At the threshold: The developing adolescent* (pp. 16–53). Cambridge, MA: Harvard University Press.
Brooks-Gunn, J., & Warren, M. P. (1989). Biological and social contributions to negative affect in young adolescent girls. *Child Development, 60,* 40–55.
Buchholz, T. G. (1984). *Margaret Mead and Samoa: The making and unmaking of an anthropological myth* [review]. *Commentary, 77,* 78–80.
Calkins, F. G. (1962). *My Samoan chief.* Honolulu: University of Hawaii Press.
Cassidy, R. (1982). *Margaret Mead: A voice for the century.* New York: Universe Books.
Caton, H. (1984, March). Margaret Mead and Samoa: In support of Freeman's critique. *Quadrant,* 28–32.
Caton, H. (Ed.). (1990). *The Samoa reader: Anthropologists take stock.* Lanham, MD: University Press of America.
Chapman, J. W., & Nichols, J. G. (1976). Occupational identity status, occupational preference, and field dependence in Maori and Pakeha boys. *Journal of Cross-Cultural Psychology, 7,* 61–72.
Coleman, J. S. (1974). *Youth: Transition to adulthood.* Chicago: University of Chicago Press.
Collins, R. (1979). *The credential society: A historical sociology of education and stratification.* New York: Academic Press.
Column Seven. (1991, January 11). *Samoa Times,* p. 2.
Côté, J. E. (1986). Identity crisis modality: A technique for measuring the structure of the identity crisis. *Journal of Adolescence, 9,* 321–325.
Côté, J. E. (1992). Was Mead wrong about coming of age in Samoa? An analysis of the Mead/Freeman controversy for scholars of adolescence and human development. *Journal of Youth and Adolescence, 21,* 499–527.
Côté, J. E. (1993). Foundations of a psychoanalytic social psychology: Neo-Eriksonian propositions regarding the relationship between psychic structure and cultural institutions. *Developmental Review, 13,* 31–53.
Côté, J. E., & Levine, C. (1987). A formulation of Erikson's theory of ego identity formation. *Developmental Review, 9,* 273–325.
Côté, J. E., & Levine, C. (1988a). A critical examination of the ego identity status paradigm. *Developmental Review, 9,* 147–184.

Côté, J. E., & Levine, C. (1988b). The relationship between ego identity status and Erikson's notions of institutionalized moratoria, value orientation state, and ego dominance. *Journal of Youth and Adolescence, 17*, 81–99.

Cox, P. A., & Elmqvist, T. (1991). Indigenous control of tropical rainforest reserves: An alternative strategy for conservation. *Ambio, 20*, 317–321.

Cranberg, L. (1983a). Ta'u revisited. *Human Organization, 42*, 182.

Cranberg, L. (1983b, July). Samoa [Letter to the editor]. *Life*, p. 19.

Crocombe, R. (1989). *The South Pacific: An introduction* (5th ed.). Suva, Fiji: University of the South Pacific.

Demos, J., & Demos, V. (1969, November). Adolescence in historical perspective. *Journal of Marriage and The Family*, 632–638.

Department of Economic Development, Western Samoa. (1987). *Western Samoa's Sixth Development Plan: 1988–1990*. Apia, Western Samoa: Author.

Department of Statistics, Western Samoa. (1989). *Annual Statistical Abstract 1989*. Apia, Western Samoa: Author.

Douglas, N., & Douglas, N. G. (1986). *Pacific Islands yearbook* (15th ed.). North Ryde, NSW, Australia: Angus & Robertson.

Drug abuse hits schools. (1991, November). *Pacific Island Monthly*, p. 23.

Education Director backs new exam system. (1991, January 30). *Samoa Observer*, p. 1.

Ember, M. (1985). Evidence and science in ethnography: Reflections on the Mead-Freeman controversy. *American Anthropologist, 87*, 906–910.

Erikson, E. H. (1956). The problem of ego identity. *Journal of the American Psychoanalytic Association, 4*, 56–122.

Erikson, E. H. (1958). *Young man Luther*. New York: Norton.

Erikson, E. H. (1963). *Childhood and society* (2nd ed.). New York: Norton.

Erikson, E. H. (1968). *Identity: Youth and crisis*. New York: Norton.

Erikson, E. H. (1980). *Identity and the life cycle: A reissue*. New York: Norton.

Erikson, K. (1966). *Wayward puritans: A study in the sociology of deviance*. New York: Wiley.

Feinberg, R. (1988). Margaret Mead and Samoa: Coming of Age in fact and fiction. *American Anthropologist, 90*, 656–663.

Festinger, L. (1957). *A theory of cognitive dissonance*. New York: Harper & Row.

Filoiali'i, L. A., & Knowles, K. (1982). The ifoga: The Samoan practice of seeking forgiveness for criminal behaviour. *Oceania, 53*, 384–388.

Flay, B. R., Bull, P. E., & Tamahori, J. (1976). Designing a questionnaire for Polynesian and Pakeha car assembly workers. *Journal of Cross-Cultural Psychology, 7*, 235–241.

Freeman, D. (1983). *Margaret Mead and Samoa: The making and unmaking of an anthropological myth*. Cambridge, MA: Harvard University Press.

Freeman, D. (1985). A reply to Ember's reflections on the Freeman-Mead controversy. *American Anthropologist, 87*, 910–917.

Freeman, D. (1987a). Comment on Holmes's quest for the real Samoa. *American Anthropologist, 89*, 930–935.

Freeman, D. (1987b). Holmes, Lowell D.: *Quest for the real Samoa: The Mead/Freeman controversy and beyond* [review]. *Journal of the Polynesian Society, 96*(3), 392–395.

Freeman, D. (1987c). Towards an anthropology both scientific and humanistic. *Canberra Anthropology*, 44–69.

Freeman, D. (1989a). Fa'apua'a Fa'amu and Margaret Mead. *American Anthropologist, 91*, 1017–1022.

Freeman, D. (1989b). Holmes, Mead, and Samoa. *American Anthropologist, 91*, 758–762.

Freeman, D. (1991). There's tricks i' th' world: An historical analysis of the Samoan researches of Margaret Mead. *Visual Anthropology Review, 7*, 103–128.

REFERENCES

Freeman, D. (1992). Paradigms in collision: The far-reaching controversy over the Samoan researches of Margaret Mead and its significance for the human sciences. *Academic Questions,* 5(3), 23–33.
Gailey, C. W. (1987). *Kinship to kingship: Gender hierarchy and state formation in the Tongan Islands.* Austin: University of Texas Press.
Gerber, E. R. (1975). *The cultural patterning of emotions in Samoa.* Unpublished doctoral dissertation, University of California, San Diego.
Gilson, R. P. (1970). *Samoa 1830 to 1900.* London: Oxford University Press.
Gold, M. G., & Petronio, R. J. (1980). Delinquent behavior in adolescence. In J. Adelson (Ed.), *Handbook of adolescent psychology* (pp. 495–535). New York: Wiley.
Goodale, J. (1984). *Margaret Mead and Samoa: The making and unmaking of an anthropological myth* [review]. *Pacific Affairs,* 57, 180–182.
Goodman, R. A. (1983). *Mead's Coming of Age in Samoa: A dissenting view.* Oakland, CA: Pipperline Press.
Gould, S. J. (1981). *The mismeasure of man.* New York: Norton.
Grattan, F. J. H. (1948). *An introduction to Samoan culture.* Papakura, New Zealand: R. McMillan.
Greksa, L. P., Pelletier, D. L., & Gage, T. B. (1986). Work in contemporary and traditional Samoa. In P. T. Baker, J. M. Hanna, & T. S. Baker (Eds.), *The changing Samoans: Behavior and health in transition* (pp. 297–326). New York: Oxford University Press.
Handler, R. (1984). Review essay/Ruth Benedict, Margaret Mead, and the growth of American anthropology. *The Journal of American History,* 71, 364–368.
Hanson, F. A. (1973). Political change in Tahiti and Samoa: An exercise in experimental anthropology. *Ethos,* 12, 1–12.
Harris, M. (1983). Margaret and the giant-killer: It doesn't matter a whit who's right. *The Sciences,* 23(4), 18–21.
Heimans, F. (producer). (1988). *Margaret Mead and Samoa* [film]. New York: Brighton Video.
Holmes, L. D. (1957a). *A restudy of Manu'a culture: A problem of methodology.* Unpublished doctoral dissertation, Northwestern University, Chicago, IL.
Holmes, L. D. (1957b). Ta'u: Stability and change in a Samoan village. *Polynesian Society Journal,* 66, 301–338, 398–435.
Holmes, L. D. (1974). *Samoan village.* New York: Holt, Rinehart & Winston.
Holmes, L. D. (1980a). Factors contributing to the cultural stability of Samoa. *Anthropological Quarterly,* 53, 188–197.
Holmes, L. D. (1980b). Cults, cargo and Christianity: Samoan responses to Western religion. *Missiology: An International Review,* 8, 471–487.
Holmes, L. D. (1983a). Margaret Mead's Samoa: Views and reviews. *The Quarterly Review of Biology,* 58, 539–544.
Holmes, L. D. (1983b). South seas squall: Derek Freeman's long-nurtured, ill-natured attack on Margaret Mead. *The Sciences,* 23, 14–18.
Holmes, L. D. (1983c). On the questioning of as many as six impossible things about Freeman's Samoa before breakfast. *Canberra Anthropology,* 6, 1–16.
Holmes, L. D. (1987). *Quest for the real Samoa: The Mead/Freeman controversy and beyond.* South Hadley, MA: Bergin & Garvey.
Holmes, L. D., & Holmes, E. R. (1992). *Samoan village then and now* (2nd ed.). Fort Worth: Harcourt Brace Jovanovich.
Hooper, A. (1984). *Margaret Mead and Samoa: The making and unmaking of an anthropological myth* [review]. *Oceania,* 55, 224–225.
Howard, A., & Kirkpatrick, J. (1989). Social organization. In A. Howard & R. Borofsky (Eds.), *Developments in Polynesian ethnology* (pp. 47–93). Honolulu: University of Hawaii Press.

Hunt, T. L., & Kirch, P. V. (1988). An archaeological survey of the Manu'a Islands, American Samoa. *Polynesian Society Journal, 97*, 153–183.
Immigration Division, Department of Labour, New Zealand. (1982). *Samoan Guide to Permanent Entry into New Zealand*. New Zealand: Author.
Irwin, G. (1965). *Samoa: A teacher's tale*. London: Cassell.
Karlsen, C. F. (1987). *The devil in the shape of a woman: Witchcraft in colonial New England*. New York: Norton.
Katchadourian, H. (1977). *The biology of adolescence*. San Francisco: Freeman.
Keesing, F. (1934). *Modern Samoa*. London: Allen & Unwin.
King, J. (1981, July/August). The myth of suicide in American Samoa. *The New Pacific Magazine*, pp. 15–18.
Klein, H. (1990). Adolescence, youth, and young adulthood: Rethinking current conceptualizations of life stage. *Youth & Society, 21*, 446–471.
Krauze, T., & Slomczynski, K. M. (1985). How far to meritocracy? Empirical tests of a controversial thesis. *Social Forces, 63*, 623–642.
Kuklick, H. (1984). Margaret Mead and Samoa: The making and unmaking of an anthropological myth [review]. *Contemporary Sociology, 13*, 558–562.
Labour Department, Western Samoa. (1984). *Labour Department Annual Report: 1984*. Apia, Western Samoa: Author.
Laing, P. K. (1987). Holmes, Lowell D.: Quest for the real Samoa:The Mead/Freeman controversy and beyond [review]. *Journal of the Polynesian Society, 96*, 395–399.
Leach, E. (1983). The Shangri-la that never was. *New Society, 63*, 477–478.
Leacock, E. (1987). Postscript: The problems of youth in contemporary Samoa. In L. D. Holmes (Ed.), *Quest for the real Samoa: The Mead/Freeman controversy and beyond* (pp. 177–188). South Hadley, MA: Bergin & Garvey.
Leacock, E. (1988). Anthropologists in search of a culture: Margaret Mead, Derek Freeman, and all the rest of us. *Central Issues in Anthropology, 8*, 3–23.
Levy, R. I. (1984). Mead, Freeman, and Samoa: The problem of seeing things as they are. *Ethos, 12*, 85–92.
Linnekin, J. (1991). Fine mats and money: Contending exchange paradigms in colonial Samoa. *Anthropological Quarterly, 64*, 1–12.
Lockhart, A. (1971). Graduate unemployment and the myth of human capital. In D. T Davies & K. Herman (Eds.), *Social space: Canadian perspective*. Toronto: New Press.
Lockhart, A. (1975). Future failure: The unanticipated consequences of educational planning. In R. M. Pike & E. Zureik (Eds.), *Socialization and values in Canadian society* (pp. 196–215). Toronto: McClelland-Stewart.
Lockwood, B. (1971). *Samoan village economy*. Melbourne: Oxford University Press.
Lockwood, V. S. (1993). *Tahitian transformations: Gender & capitalist development in a rural society*. Boulder, CO: Lynne Rienner.
MacPherson, C. (1988). The road to power is a chainsaw: Villages and innovation in Western Samoa. *Pacific Studies, 11*, 1–24.
MacPherson, C., & MacPherson, L. (1985). Suicide in Western Samoa: A sociological perspective. In F. X. Hezel, D. H. Rubenstein, & G. M. White (Eds.), *Culture, youth and suicide in the Pacific: Papers from the East-West Center Conference* (pp. 36–73). Honolulu: East-West Center.
Mageo, J. M. (1988). Malosi: A psychological exploration of Mead's and Freeman's work and of Samoan agression. *Pacific Studies, 11*, 25–65.
Mageo, J. M. (1989). Aga, Amio and Loto: Perspectives on the structure of the self in Samoa. *Oceania, 59*, 181–199.
Malifa, S. (1975). *Looking down at waves: A collection of poems*. Suva, Fiji: Mana Publications.
Marcia, J. E. (1964). *Determination and construct validation of ego identity status*. Unpublished doctoral dissertation, Ohio State University, Columbus.

REFERENCES

Marcia, J. E. (1976). *Studies in ego identity.* Unpublished manuscript.
Marcia, J. E. (1980). Identity in adolescence. In J. Adelson (Ed.), *Handbook of adolescent psychology* (pp. 159–187). New York: Wiley.
Marquardt, C. (1984). *The tattooing of both sexes in Samoa* (S. Ferner, Trans.). Papakura, New Zealand: R. McMillan. (Original work published 1899)
Maugham, W. S. (1985). *The trembling of a leaf.* Honolulu, HI: Mutual Publishing.
McCabe, B. (1992, April). Samoan sights on new business. *Pacific Islands Monthly,* p. 39.
McDowell, N. (1984). Review of *Margaret Mead and Samoa. Pacific Studies, 4,* 99–140.
McGovern, B. (1988). *Western Samoa.* Sydney, Australia: South Pacific Trade Commission.
Mead, M. (1928). *Coming of age in Samoa: A psychological study of primitive youth for Western Civilization.* New York: Morrow Quill Paperbacks.
Mead, M. (1931). The role of the individual in Samoan culture. In A. L. Kroeber & T. T. Waterman (Eds.), *Source book in Anthropology* (pp. 545–561). New York: Harcourt, Brace & World.
Mead, M. (1937). A reply to a review of "sex and temperament in three primitive societies." *American Anthropologist, 39,* 558–561.
Mead, M. (1961). [Review of] *Ta'u: Stability and change in a Samoan village,* by Lowell D. Holmes. *American Anthropologist, 63,* 428–430.
Mead, M. (1969). *Social organization of Manu'a* (2nd ed.). Honolulu: Bernice P. Bishop Museum. (Original work published 1930)
Mead, M. (1972). *Blackberry winter: My earlier years.* New York: Morrow.
Mead, M. (1977). *Letters from the field, 1925–1975.* New York: Harper & Row.
Meleisea, M. (1987a). *Lagaga: A short history of Western Samoa.* Suva, Fiji: University of South Pacific.
Meleisea, M. (1987b). *The making of modern Samoa: Traditional authority and colonial administration in the modern history of Western Samoa.* Suva, Fiji: Institute of Pacific Studies of the University of the South Pacific.
Meleisea, M., & Meleisea, P. (1980). "The best kept secret": Tourism in Western Samoa. In Institute of Pacific Studies (Eds.), *Pacific tourism: As islanders see it.* Suva, Fiji: University of the South Pacific.
Montemayor, R., Adams, G. R., & Gullotta, T. P. (1990). *From childhood to adolescence: A transitional period?* Newbury Park, CA: Sage.
Montemayor, R., & Flannery, D. J. (1990). Making the transition from childhood to early adolescence. In R. Montemayor, G. R. Adams, & T. P. Gullotta (Eds.), *From childhood to adolescence: A transitional period?* (pp.). Newbury Park, CA: Sage.
Moyle, R. (Ed.). (1984). *The Samoan journals of John Williams 1830–1832.* Canberra: Australian National University Press.
Muensterberger, W. (1985). *Margaret Mead and Samoa: The making and unmaking of an anthropological myth* [review]. *Psychoanalytic Quarterly, 54,* 101–105.
Murray, S. O. (1990). Problematic aspects of Freeman's account of Boasian culture. *Current Anthropology, 31,* 401–407.
Murray, S. O. (1991). On Boasians and Margaret Mead: Reply to Freeman. *Current Anthropology, 32,* 448–452.
Muse, C. J. (1991). Women in Western Samoa. In L. A. Adler (Ed.), *Women in cross-cultural perspective* (pp. 221–240). New York: Praeger.
Muuss, R. E. (1988). *Theories of adolescence* (5th ed.). New York: Random House.
Nardi, B. A. (1984). The height of her powers: Margaret Mead's Samoa. *Feminist Studies, 10,* 323–337.
Natarajan, N. (1983). *Rural development projects and women in Western Samoa. ILO study.* Apia, Western Samoa: University of South Pacific.
Ngan-Woo, F. E. (1985). *FaaSamoa: The world of Samoans.* Auckland, New Zealand: Office of the Race Relations Conciliator.

North, D. (1991, February). American Samoans doing the most. *Pacific Islands Monthly*, p. 25.
North, D. (1993, April). Sputtering unions. *Pacific Islands Monthly*, p. 12.
Norton, R. (1984). Titles, wealth and faction: Electoral politics in a Samoan village. *Oceania, 55*, 100–117.
O'Meara, T. (1990). *Samoan planters: Tradition and economic development in Polynesia*. Fort Worth, TX: Holt, Rinehart & Winston.
Offer, D., & Offer, J. B. (1975). *From teenage to young manhood: A psychological study*. New York: Basic Books.
Oliver, D. (1961). *The Pacific Islands*. New York: Doubleday.
Paikoff, R. L., & Brooks-Gunn, J. (1991). Do parent–child relationships change during puberty? *Psychological Bulletin, 110*, 47–66.
Park, C. B. (1980). *The population of American Samoa*. Honolulu: University of Hawaii School of Public Health and East-West Population Institute.
Patience, A., & Smith, J. W. (1986). Derek Freeman and Samoa: The making and unmaking of a biobehavioral myth. *American Anthropologist, 88*, 157–162.
Paxman, D. B. (1988). Freeman, Mead, and the eighteenth-century controversy over Polynesian society. *Pacific Studies, 11*, 1–19.
Petaia, R. (1980). *Blue rain*. Apia, Samoa: U.S.P Centre, Western Samoa.
Pfieffer, M., & Côté, J. E. (1991). Inglehart's silent revolution thesis: An examination of life-cycle effects in the acquisition of postmaterialist values. *Social Behavior and Personality, 19*, 223–235.
Phillips, D. P. (1979). Suicide, motor vehicle fatalities and the mass media: Evidence toward a theory of suggestion. *Sociological Review, 84*, 1150–1174.
Pitt, D. (1970). *Tradition and economic progress in Samoa: A case study of the role of traditional social institutions in economic development*. Oxford: Clarendon Press.
Pitt, D., & MacPherson, C. (1974). *Emerging pluralism: The Samoan community in New Zealand*. Auckland, New Zealand: Longman Paul.
Police & Prisons Department, Western Samoa (1965). *Annual Report of the Police & Prisons Department—1965*. Apia, Western Samoa: Author.
Popper, K. (1968). *The logic of scientific discovery*. New York: Harper & Row.
Porter, J. (1984). Education and the just society. In A. Himelfarb & C. J. Richardson (Eds.), *Sociology for Canadians: A reader* (pp. 460–473). Toronto: McGraw-Hill Ryerson.
Rappaport, R. A. (1986). Desecrating the holy woman: Derek Freeman's attack on Margaret Mead. *American Scholar, 55*(3), 313–347.
Rappaport, R. A. (1987a). Reply to Freeman. *American Scholar, 56*(2), 304.
Rappaport, R. A. (1987b). Reply to Freeman. *Scientific American, 256*, 6–7.
Reid, B. V. (1990). Weighing up the factors: Moral reasoning and culture change in a Samoan community. *Ethos, 18*, 48–70.
Reyman, J. E. (1985). Some comments on the Freeman-Mead controversy. *American Anthropologist, 87*, 393–394.
Robie, D. (1991, January). Troubled times reflected in Samoans' inner search [film review]. *Pacific Islands Monthly*, pp. 45–46.
Rokeach, M. (1960). *The open and closed mind*. New York: Basic.
Rowe, N. A. (1930). *Samoa under the sailing gods*. London: Putnam.
Rubinstein, D. H. (1985). Suicide in Micronesia. In F. K. Hezel, D. H. Rubenstein, & G. H. White (Eds.), *Culture, youth and suicide in the Pacific: Papers from an East-West Center Conference* (pp. 88–111). Honolulu: Center for Asian and Pacific Studies, University of Hawaii at Manoa.
Salmi, J. (1992). The higher education crisis in developing countries: Issues, problems, constraints and reforms. *International Review of Education, 38*, 19–33.

REFERENCES

Scheper-Hughes, N. (1984). The Margaret Mead controversy: Culture, biology, and anthropological inquiry. *Human Organization, 43,* 85-93.
Schoeffel, P. (1979). *Daughters of Sina: A study of gender, status and power in Western Samoa.* Unpublished doctoral dissertation, Australian National University, Canberra.
Schoeffel, P. (1986). The Samoan journals of John Williams 1830-1832 [review]. *Oceania, 57,* 63-64.
Schoeffel, P., & Meleisea, M. (1983). Margaret Mead, Derek Freeman and Samoa: The making, unmaking and remaking of an anthropolical myth. *Canberra Anthropology, 6,* 58-69.
Seligman, M. E. P. (1988, October). Boomer blues. *Psychology Today,* pp. 50-55.
Shankman, P. (1983). The Samoan conundrum. *Canberra Anthropology, 6*(1), 38-57.
Shore, B. (1982). *Sala'ilua: A Samoan mystery.* New York: Columbia University Press.
Shore, B. (1983). Paradox regained: Freeman's Margaret Mead and Samoa. *American Anthropologist, 85,* 935-944.
Special youth jail? (1992, April). *Pacific Island Monthly,* p. 16.
Spindler, G., & Spindler, L. (1990). *Foreword.* In T. O'Meara, *Samoan planters: Tradition and economic development in Polynesia.* Fort Worth, TX: Holt, Rinehart & Winston.
Sprinthall, N. A., & Collins, W. A. (1984). *Adolescent psychology: A developmental view.* Reading, MA: Addison-Wesley.
Stack, S. (1987). Celebrities and suicide: A taxonomy and analysis. *American Sociological Review, 52,* 401-412.
Stair, J. B. (1983). *Old Samoa or flotsam and jetsam from the Pacific Ocean.* Papakura, New Zealand: R. McMillan. (Original work published 1897)
Stanner, W. E. H. (1953). *The south seas in transition.* Sydney: Australasian Publishing Co.
Swaney, D. (1990). *Samoa: Western & American Samoa.* Hawthorn, Victoria, Australia: Lonely Planet Publications.
Taxi license deadline extended. (1991, January 11). *Samoa Observer,* p. 1.
Tepperman, L. (1977). *Crime control: The urge toward authority.* Toronto: McGraw-Hill Ryerson.
Treasury Department of the Government of Western Samoa. (1988). *Western Samoa: Setting a new course for growth.* Institutional Investor.
Tribe, K. (1984). *Margaret Mead and Samoa: The making and unmaking of an anthropological myth* [review]. *Sociological Review, 32,* 398-401.
Turnbull, C. M. (1983, March 28). *Margaret Mead and Samoa: The making and unmaking of an anthropological myth* [review]. *The New Republic,* 32-34.
Turner, G. (1986). *Samoa: Nineteen years in Polynesia.* Apia, Samoa: Western Samoa Historical and Cultural Trust. (Original work published 1861)
Twain, M. (1924). Autobiography. New York: Harper & Brothers.
Ward, R. G. (1967). *American activities in the central Pacific 1790-1870 Vol. 6.* Ridgewood, NJ: The Gregg Press.
Wendt, A. (1973). *Sons for the return home.* London: Penguin.
Wendt, A. (1974). *Some poetry from Western Samoa.* Suva, Fiji: Mana Publications.
Wendt, A. (1977). *Pouliuli.* London: Penguin.
Wendt, A. (1979). *Leaves of the banyon tree.* London: Penguin.
Wendt, A. (1983, April). Three faces of Samoa: Mead's, Freeman's and Wendt's. *Pacific Islands Monthly,* pp. 10-14, 69.
Wendt, F. (1984). Review of Margaret Mead and Samoa. *Pacific Studies, 40,* 92-99.
Western Samoa push for investment. (1992, April). *Pacific Island Monthly,* p. 31.
Williamson, R. W. (1975). *Essays in Polynesian ethnology.* Cambridge: Cambridge University Press. (Original work published 1939)
Young, R. E., & Juan, S. (1985). Freeman's Margaret Mead myth: The ideological virginity of anthropologists. *Australia and New Zealand Journal of Sociology, 21,* 64-81.
Yusuf, S., & Peters, R. K. (1985). *Western Samoa: The Experience of Slow Growth and Resource Imbalance (World Bank Staff Working Papers).* Washington: The World Bank.

Author Index

A

Adams, G. R., 18, 151
Aiavao, U., 130, 131
Ala'ailima, F., 9, 10, 41
Allahar, A. L., 146, 147
Appell, G. N., 3, 16, 22, 23, 30, 62
Aries, P., 151
Ayabe, H. I., 11, 163

B

Badock, C. R., 23, 30
Baker, P. T., 41, 165
Baker, T. S., 30, 41, 165
Baldauf, R. B., Jr., 11, 163
Ballantine, J. H., 140, 146, 147
Barnouw, V., 102
Baumeister, R. F., 162
Benedict, R., 154
Beneteau, R., 132
Berg, I., 141
Bernard, R. H., 62
Bock, P. K., 10
Bowles, J. R., 129, 131
Brady, I., 4, 5, 7, 22, 30
Brake, M., 156
Brewer, W. K., 95
Broad, W., 20

Brooks-Gunn, J., 17, 18
Buchholz, T. G., 16, 30
Bull, P. E., 165

C

Calkins, F. G., 41
Cassidy, R., 75
Caton, H., 3, 21, 22, 26, 28, 30
Chapman, J. W., 168
Coleman, J., 151
Collins, R., 140, 141, 151
Collins, W. A., 16, 18, 150
Côté, J. E., 4, 33, 140, 160, 163, 165, 167
Cox, P. A., 146
Cranberg, L., 27, 61
Crocombe, R., 137, 146, 147, 168

D

Demos, J., 151
Demos, V., 151
Department of Economic Development, Western Samoa, 138, 139, 141, 142, 143
Department of Statistics, Western Samoa, 38, 45

AUTHOR INDEX

Douglas, N., 54, 139
Douglas, N. G., 54, 139

E

Elmqvist, T., 146
Ember, M., 30
Erikson, E. H., 10, 67, 105, 134, 152, 153, 165, 167
Erikson, K., 158

F

Feinberg, R., 10, 62, 63
Festinger, L., 29
Filoiali'i, L. A., 74
Flannery, D. J., 18
Flay, B. R., 165
Freeman, D., 1, 3, 4, 7, 14, 16, 18, 22, 23, 24, 25, 26, 27, 28, 29, 30, 31, 33, 35, 36, 38, 40, 41, 43, 44, 45, 53, 73, 75, 82, 99, 114, 115, 116, 127, 160

G

Gage, T. B., 67
Gailey, C. W., 122,
Gerber, E. R., 28, 29, 86, 158
Gilson, R. P., 73, 83
Gold, M. G., 32, 39, 151, 160
Goodale, J., 30
Goodman, R. A., 30
Gould, S. J., 20
Grattan, F. J. H., 68, 72
Greksa, L. P., 67
Gullotta, T. P., 18, 151

H

Handler, R., 30
Hanna, J. M., 31, 165
Hanson, F. A., 11, 72, 87
Harris, M., 40
Heimans, F., 3, 23, 30
Holmes, E. R., 3, 125

Holmes, L. D., 3, 5, 9, 11, 27, 36, 38, 45, 48, 49, 50, 59, 62, 66, 67, 71, 72, 74, 81, 86, 87, 89, 90, 91, 92, 93, 97, 125, 158, 168
Hooper, A., 10, 30
Howard, A., 82
Hunt, T. L., 53, 54

I

Immigration Division, New Zealand, 134
Irwin, G., 97

J

Juan, S., 32

K

Karlsen, C. F., 77
Katchadourian, H., 20, 35
Keesing, F., 90, 92
King, J., 125
Kirch, P. V., 53, 54
Kirkpatrick, J., 82
Klein, H., 151
Knowles, K., 74
Krauze, T., 146
Kuklick, H., 30

L

Labour Department, Western Samoa, 134, 139, 142
Laing, P. K., 10, 30
Leach, E., 30
Leacock, E., 30, 41, 94, 127, 129, 136, 143
Levine, C., 140, 163, 165, 167
Levy, R. I., 10, 43
Linnekin, J., 104
Lockhart, A., 140
Lockwood, B., 56
Lockwood, V. S., 93, 122

M

MacPherson, C., 41, 74, 94, 129, 130, 134, 138, 145

MacPherson, L., 41, 74, 94, 129, 130, 134
Madan, T. N., 3, 22, 23, 30
Mageo, J. M., 10, 72, 105, 167
Malifa, S., 132
Marcia, J. E., 33, 160, 165, 166, 167
Marquardt, C., 68, 69, 70
Maugham, W. S., 86
McCabe, B., 143, 144
McDowell, N., 15
McGovern, B., 89
Mead, M., 1, 2, 9, 10, 11, 15, 17, 25, 31, 34, 44, 45, 49, 58, 59, 60, 61, 63, 71, 75, 77, 81, 89, 94, 100, 103, 111, 123, 149, 154, 156, 161, 162
Meleisea, M., 5, 9, 11, 25, 28, 30, 43, 50, 51, 52, 53, 56, 57, 67, 69, 70, 73, 74, 76, 84, 85, 87, 90, 94, 99, 125, 126, 127, 128, 133, 142
Meleisea, P. (see Schoeffel, P.)
Montemayor, R., 18, 151
Moyle, R., 75, 76, 77, 78, 84, 106
Muensterberger, W., 30
Murray, S. O., 30
Muse, C. J., 142
Muuss, R. E., 19, 41, 59

N

Nardi, B. A., 30
Natarajan, N., 142
Ngan-Woo, F. E., 71, 98
Nicholls, J. G., 168
North, D., 124, 144
Norton, R., 94, 134

O

O'Meara, T., 7, 38, 41, 43, 70, 89, 93, 94, 126, 127, 128, 129, 130, 134, 144
Offer, D., 33
Offer, J. B., 33
Oliver, D., 86, 87, 122, 125

P

Park, C. B., 54
Paikoff, R. L., 17

Patience, A., 30
Paxman, D. B., 30
Pelletier, D. L., 67
Petaia, R., 132
Peters, R. K., 41, 93
Petronio, R. J., 32, 39, 151, 160
Pfeiffer, M., 33
Phillips, D. P., 131
Pitt, D., 41, 91, 138, 142, 145
Police & Prisons Department, Western Samoa, 45
Popper, K., 29
Porter, J., 140, 146

R

Rappaport, R. A., 15, 17, 18, 19, 30
Reid, B. V., 165, 167
Reiter, E. O., 17
Reyman, J. E., 30
Robie, D., 132
Rokeach, M., 22
Rowe, N. A., 44, 89, 96
Rubinstein, D. H., 132

S

Salmi, J., 142
Scheper-Hughes, N., 30
Schoeffel, P., 5, 9, 25, 28, 29, 30, 50, 51, 52, 53, 56, 57, 74, 99, 125, 126, 128, 142
Schoeffel-Meleisea, P. (see Schoeffel, P.)
Seligman, M. E. P., 163
Shankman, P., 129
Shore, B., 30, 58, 93
Slomczynski, K. M., 146
Smith, J. W., 30
Spindler, G., 38
Spindler, L., 38
Sprinthall, N. A., 16, 18, 150
Stack, S., 131
Stair, J. B., 80
Stanner, W. E. H., 11, 58, 72, 87
Swaney, D., 9, 10, 53, 60, 131

T

Tamahori, J., 165
Tepperman, L., 32, 44

AUTHOR INDEX

Treasury Department, Western Samoa, 143, 144
Tribe, K., 30
Turnbull, C. M., 8, 30, 53, 61
Turner, G., 39, 68, 70, 71, 72, 79, 83, 84, 85, 88, 89, 93, 94
Twain, M., 43

W

Wade, N., 20
Ward, R. G., 54
Warren, M. P., 18
Wendt, A., 6, 9, 10, 41, 75, 87, 115, 132, 132
Wendt, F., 9, 10, 29, 53, 58
Williamson, R. W., 80, 81, 82

Y

Young, R. E., 32
Yusuf, S., 41, 93

Subject Index

A

Adolescence, 1-2, 12, 149-152
 biology of, 14-19, 40, 151-152, 163
 cultural variation, 149-150
 definition of, 151
 female, 69-72, 103-110
 male, 68-72, 110-113
 nature of, 14, 152
 period of dependency, 124
 social history of, in precontact Samoa, 65-82
 social structure of, 39
 stage of life, 149-152
Adolescent turmoil (see storm and stress)
Aggression, 3, 39, 41
 gang violence, 125
 murder and assault, 45-46
'Aiga (extended family), 69, 74, 94, 98, 157, 158
Alienation, 132
American Anthropology "establishment," 21-22
American Samoa, 10, 12, 94, 107, 124, 125,
Apia, 37, 38, 124, 143
Aualuma (the group of unmarried women in a village), 69, 71-72, 75, 90-92, 95, 108, 111, 123, 127, 134, 137, 141, 156

'Aumaga (the group of untitled men in a village), 30, 68, 71-73, 90-95, 111, 112, 113, 123, 127, 137, 141, 153, 156, 167
 ego development and, 167

B

Benedict, Ruth, 8, 24
Boas, Franz, 3, 7, 8
Burt, Sir Cyril, 20, 36

C

Capitalism,
 cultural disenfranchisement and, 122-125
 dependency theory, 146-148
 identity and, 169
 impact of, on youth, 123-124
 international capitalism, 145-148, 165, 169
 missionaries as shocktroops of, 122
 modernization theory, 145
Choice, 31, 70-71, 112, 160-163
 depression and, 163
 deviance and, 161-162
 individualism and, 163

SUBJECT INDEX

Christianized Samoa, 6–7, 11,
Coming of age, 151–152
   in contemporary Samoa, 122–148
   in Mead's Samoa, 2, 102–121, 160
   in precontact Samoa, 66–74
   in Western Samoa, 125–136
   missionary impact on, 87–95
   preparation for adulthood and, 105
   quid pro quo in, 66–67, 124, 128,
Context of Mead's research 1–3
Crime rates, 43–46
Cultural determinism, 20,
Cultural disenfranchisement, 39, 41, 88, 95, 122–125, 137
Cultural stability, 72–73, 155
   role of coming of age in, 72

D

Delinquency rates, 31–41
Delinquents, 30–31, 34, 113
Deviance, 30–35, 55, 73–74, 158–160
Deviants,
   downward, 30–35
   upward, 30–31, 33, 106, 109, 161
Division of labor, 52, 69–71, 108–109, 135, 157

E

Education, 88–89, 103, 105, 107, 128, 138–142, 151, 154
   brain drain, 146
   credentialism, 141
   cultural capital, 146
   financing of, 139
   human capital theory, 140, 142, 145
   mass education, 92, 94, 95, 106, 122, 151
   National University of Samoa, 139
   policy, 138–140
   programs, 139–142
   Samoanization of, 138
Ego development, 165–168
   ego mastery, 167

F

Fa'apua'a Fa'amu, testimony of, 25–29
Fa'a Samoa (the Samoan way), 11, 87, 95, 125, 141, 145

Females,
   adolescence in Mead's Samoa, 103–110, 156
   decline in status, 91–92, 123, 134–136
   entry into adulthood, 108–109
   roles and status, in precontact Samoa, 69–72
Fine mats, 69, 74, 104–105
Fono (village assembly of *matai*), 5, 6, 87, 93, 157
   ego mastery and, 167
   postconventional reasoning and, 165
Freeman, Derek,
   anecdotal evidence, use of, 40
   case against Mead, 14–47
   claim to have resolved the controversy, 29–30
   criticisms of, in the literature, 4, 30, 49, 51–52, 115, 129
   denial of relevance of social change, 6–7, 36–38, 41, 53, 58, 65–66, 129, 136
   denial of Samoa's non-Christian past, 123
   followers, 3, 21–26
   methodological vagueness, 38–39
   "mission" of, 6, 82, 116, 136
   reaction to criticism, 4, 22–23
   "refutation," 3, 6, 35,
   writing and referencing style, 15

G

Gender roles, 123, 134–136

H

Human development, 19
   theory, 39–41
Hurricanes, 67

I

Identity, 152, 159–160, 165–169
   authority and, 169
   capitalism and, 169
   culture and, 169
   ego identity, 166–167

Identity *(cont.)*
    identity crisis, 165–166
    identity status paradigm, 165–168
    cross-cultural relevance, 166–168
Ifoga (forgiveness ceremony), 74, 87, 157
Individualism, 84–85, 125, 141
    depression and, 163
Institutionalized moratorium (see psychosocial moratorium)
Interactionism, 21, 34, 41

L

London Missionary Society (LMS), 83–84

M

Malaga (visiting party), 90, 91, 95, 153
Males, 123
    adolescence in Mead's Samoa, 110–113
    learning of cooperation in groups, 111–112
    roles and status in precontact Samoa, 68–72
Manu'a, 11, 54, 56, 67, 89, 97, 111, 124
    map of, 37
Margaret Mead and Samoa, the film, 23–30
Matai (a titled family head, or chief), 5–7, 10, 69, 71, 72, 73–74, 91, 93, 94, 95, 108, 111, 113, 127, 128, 129, 130, 155, 156, 157
    service to (*tautua*), 69, 73, 93, 95, 128, 153, 157
Maugham, Somerset, 86
Mead–Freeman controversy, 12, 19, 58, 66, 69, 75, 87, 102, 122, 134, 136, 137, 149, 150, 164, 165
    context of, 3–4
    Freeman as prosecutor, 5–8
    Mead as defendant, 8–11
    nature-nurture debate and, 20–23
    politics of, 19–30
Mead, Margaret
    account of premarital sexual behavior, 96, 98, 114–121
    coming-of-age thesis, 19, 25, 33, 49–50, 58, 61–63, 75, 90, 100, 121, 129, 164

conclusions, 2, 14, 15, 20, 49, 109
culpability in the controversy, 59–60
dispute with Lowell Holmes, 49, 59
generalizations, 10, 49, 58, 63, 102–103, 118
Lowell Holmes' view of, 48–50
methodology of her study, 2, 15–18, 62, 103
mythology surrounding her book, 34, 75, 114, 118, 121
own defense, 58–61
predictions of changes in Samoan culture, 94, 106, 119, 123
qualifications placed by her upon her findings, 100–103
residence during her study, 60–61
Schoeffel and Meleisea's view of, 50–52
supposed contradictions in her study, 30–35
verdict regarding her thesis, 62–64
Mead's Samoa, 100–121, 160
    adolescence in, 104–109, 111–113
    childhood in, 103–104, 110–111, 113
    compared with contemporary Western Samoa, 127
    institutionalized moratoria and, 152–160
        deviance, 158–160
        role induction, 156
        social control mechanisms, 157
        social organization, 154–155
    sexual behavior in, 109–110, 114–121
    shame-orientation of, 158
    transitional nature of, 101–103, 114
Methodology, 11–13
    limits of, in this study, 164
    future research, 165
Missionaries, 6–7, 66, 83–99, 100, 103, 110, 158
    impact on Samoan culture, 83–99, 137
    impact on sexual behavior, 116–118
    patriarchal attitude, 91, 123, 134
    plan to change Samoa, 72, 84–86, 88, 93–94, 107, 125
    shocktroops of capitalism, 85, 122
    Turner, 68, 79, 83–85, 88–89
    Williams, 75–78, 82, 83, 106
    verification of Mead's account, 77, 106
Moratorium period (see Psychosocial moratorium)

SUBJECT INDEX

N

Nature–nurture debate, 1, 3, 5, 7, 12, 19, 20, 23
New Zealand, 94–95, 98, 124, 126, 133–134, 138, 146, 168

P

Palagi (a White person, European), 90, 94, 97, 141, 146, 148
Papalagi (see *palagi*)
Paraquat, 131
Pastors, 6, 82, 93, 103, 105, 107, 137
  boarding school/house, 103, 106, 107, 108, 116–117, 161
Personality tests, 50
Policy recommendations, 136–148
  economic, 142–145
  education, 138–142
  independence from economic domination, 145–148
  religion, 137–138
Popper, Karl, 7, 29
Premarital chastity (see virginity)
Psychosocial moratorium, 10, 63, 67, 105, 107, 108, 109, 112, 134, 152–160
  characteristics of effective, 153
    deviance and, 158
    role induction in, 155–156
    social control in, 157
    social organization of, 154
Puberty, 1–2, 17–18, 33–34, 39, 40, 104, 117

R

Race, 16
Rape, 3, 43–45
Rites of passage, 39, 68–72, 95, 111

S

Samoa,
  geography, 53–56
  history, 56–58

religion, 96, 106, 126, 137–138, 160–161
Samoan character, 60
Samoan culture, 9–11, 36, 38, 49, 65, 87
  changes in,
    childhood, 52
    Mead's account of, 101–103, 114
    resistance to Western influence, 86
    sexual practices, 91–92, 95–99
    sociosexual conditions, 95–98
  Christianization of, 83–99
  dualism of, 50–52, 86
  missionary attitude toward, 83–87
  precontact, 66–82
    affluence of, 67
    childhood in, 66–67, 150
    conceptions of guilt, 86
    sexual practices in, 74–82,
    shame-based nature of, 86
    tolerance of premarital sex in, 77
Samoan honor, 6, 19, 116
Samoan Reader, The, 22
Sexual behavior, 3, 24–29, 51, 74–82, 91–92, 95–99, 109–110, 115–121
  clandestine, 105, 127
  discretion required, 107, 110
  in contemporary Western Samoa, 135
  Mead's account compared with Freeman's, 114–121
  passive acceptance of premarital, 103, 107
  social organization of in Mead's Samoa, 120–121
Social control mechanisms, 41, 157–158
Social disorganization, 40, 41, 142
Social Organization of Manu'a, 9, 58–60, 102
Stanley Hall, G., 1
Statistical arguments, misleading, 31–33, 43–46
Storm and stress, 1–3, 14, 20, 30, 46, 63, 65, 103, 149, 157, 159–160
Suicide, 125, 129–132, 136

T

Tattooing, 39. 88–90
  of men, 68–69, 70, 88–90, 111, 113
  of women, 69–71, 89–90

Ta'u, 3, 10, 27, 31, 36, 44, 46, 53–57, 67, 74, 84, 97, 100, 108, 111, 121, 149
  map of, 37
Taupou, 76, 80, 82, 90, 92, 96, 108, 118
Temperament, 34, 158, 162
Theft, 38, 126

V

Virginity, 3, 96, 98–99, 115
  Mead's Samoa, 103, 106, 107, 115–118
  precontact Samoa, 79–81

W

Wage labor, 85, 92–93, 94, 123, 125, 128, 130, 141, 142–145
Western impact (influence) on Samoa, 41, 83–99, 102–103, 122–125, 136
  dealing with, in contemporary Western Samoa, 136–148
Western Samoa, 5, 6, 12, 36, 38, 43, 94, 122, 124, 125–148
  emigration from, 133
  map of, 37
  'Upolu, 53–54, 124
Women's Committee, 90–91